CANOE COUNTRY

Wildlife

A Field Guide to the
Boundary Waters and Quetico

CANOE COUNTRY

A Field Guide to the
Boundary Waters and Quetico

Mark Stensaas

Illustrations by Rick Kollath

Pfeifer-Hamilton

Pfeifer-Hamilton Publishers
210 West Michigan
Duluth MN 55802-1908 218-727-0500

Canoe Country Wildlife
A Field Guide to the Boundary Waters and Quetico

Printed in the United States of America by Edwards Brothers Incorporated
10 9 8 7 6 5 4 3 2 1

Editorial Director: Susan Gustafson
Assistant Editor: Patrick Gross
Art Director: Joy Morgan Dey

Library of Congress Cataloging in Publication Data
92-70991

ISBN 0-938586-65-3

Dedicated to the one Lord God who created this world
and divinely designed all its fabulous creatures

CONTENTS

Reptiles and Amphibians

Insects and Other Invertebrates

Appendix

Acknowledgments

I'll never forget our Saturday hikes. I was in high school then, and birding friends were hard to come by. My "wandering buddy" Chris Anderson, his uncle Jerry, and I would tramp all day up, over, across, and through the wild lands of central Minnesota's Hennepin County, stopping to check out whatever caught our eye, be it a soaring Red-tailed Hawk, a deer skull, or a mass of Leopard Frogs. Chris and I were going to be wildlife biologists when we grew up.

Eastman Nature Center naturalist Kathy Sandell became our mentor, showing us the ropes of being an outdoor educator. Her caring influence made an indelible mark on my life.

Time spent in the woods with friends and naturalists is always enjoyable and is often a great learning experience. As they shared their knowledge of the natural world, their exuberant curiosity rubbed off on me. A big thanks to Chris Anderson, "Uncle" Jerry, my uncle Ben Thompson, Dave Swanovich, Kathy Sandell, Ken Gilbertson, Don Stauty, Jill Laughlin, Tim Conklin, Peter Olson, Al Hagen, Dan Burg, Gary Irons, Jon Hinkel, Paul Webster, Chris Evavold, Bunter Knowles, Kim Eckert, D. Parker Backstrom, Jim Wiinanen, Brett Nelson, Tammy Gravning, and Timo Rova.

Much of the scientific data in this book was gleaned from current scientific journals. Without the years of field work and lab research by dedicated and perceptive ornithologists, mammalogists, herpetologists, ichthyologists, entomologists, ecologists, physiologists, ethologists, botanists, cell biologists, chemists and other scientists, this book would have been impossible.

My scientific reviewers thoroughly perused the text to make sure all the facts were in order and accurate. Thanks to Bunter Knowles and Cal Harth for going over the entire manuscript with a fine-toothed comb. I'm also indebted to Dr. Don Christian for reviewing the mammal section, Pat Collins and Dr. Gerald Niemi who checked the bird chapters, and Larry Webber for double checking all the insect and invertebrate facts.

Rick Kollath's excellent pen-and-ink illustrations beautifully bring to life the canoe country's critters and add flavor to the text.

Pfeifer-Hamilton Publishers is full of good and talented people. I'm not sure this project would have been completed without Susan Gustafson's guidance, after-hours effort, and encouragement when my enthusiasm was at a low ebb. Patrick Gross also went above and beyond in his work on the manuscript to make the book a success. Joy Dey is a talented graphic artist who designed the marvelous layout and cover for this book.

Editor Carol Kennedy somehow managed to make sense out of my jumbled writing and transform it into readable text.

I'm very grateful to my publisher Don Tubesing for giving me the opportunity to write *Canoe Country Wildlife* and also for publishing great books about the Arrowhead region of Minnesota, which is dear to many of us.

And, finally, thanks to my parents for their support, allowing me the freedom to take the "road less travelled."

"Sparky"
Mark Stensaas
September 1992

Introduction

The Quetico-Superior wilderness, a 16,000-square-mile canoe-ists' paradise of interconnecting lakes, lies at the very heart of the North American continent and straddles the United States–Canadian border.

Declared a wilderness area in 1964, Minnesota's Boundary Waters Canoe Area Wilderness (BWCAW) is the largest U.S. wilderness east of the Rocky Mountains and the most heavily used portion of the entire U.S. wilderness system. Its twenty-five hundred lakes, many connected to one another by overland portages, create an amazing 1200-mile labyrinth of mapped canoe routes.

Ontario's Quetico Provincial Park was granted wilderness status and protection in 1973. The Quetico's forests are also laced with a myriad of clear blue lakes and small rivers. Lake Saganaga, Knife Lake, Basswood, and Lac la Croix form the 100-mile international border between the Boundary Waters and Quetico—between the United States and Canada.

The Boundary Waters and Quetico wilderness is home to hundreds of wildlife species. Visited in their natural surroundings on their home turf, these friendly critters add immensely to the adventure of traveling through this unspoiled natural retreat. Unfortunately, many people, even those who love the area and visit it regularly, know little about the creatures who live there. They hurry across the portages and paddle across the lakes so quickly that they miss a good share of the intrigue and the beauty of the wildlife around them.

This friendly field guide

Canoe Country Wildlife is a field guide to the most commonly encountered mammals, birds, reptiles, amphibians, fish, insects and other invertebrates native to the Boundary Waters Canoe Area Wilderness and the Quetico Provincial Park. It is designed primarily for spring, summer, and fall visitors, be they canoe trippers, day paddlers, anglers, guides, lodge guests, or local residents.

The joy of wildlife watching grows when you get to know the habits and understand the lifestyles of each creature that you meet. But when you're out there how do you correctly identify the wildlife that you see? Or notice their habits? Or understand their lifestyles? *Canoe Country Wildlife* helps you answer these questions. It helps you to see more—and to understand what you are seeing.

The 75 species of wildlife included in this book are those you are most likely to see or hear on your canoe trip into the Boundary Waters and Quetico. However, even if you're a most observant paddler, you won't see them all on a single trip. And because this wilderness is home to many more species than are included in this book, you may encounter a number of creatures not described here. For quick reference, complete checklists of all Boundary Waters and Quetico mammals, birds, reptiles, amphibians, and fish are found at the end of their respective sections—not insects, though, there are far too many.

The format

To facilitate easy reference, the information for each species is presented in a consistent format.

Quick identifiers—Accurate line drawings and the brief descriptions that accompany them will help you rapidly identify the wildlife you see.

Natural history—Information about the physical characteristics, habitat, food preferences, and daily activities of each animal will help you know where, when, and how you might encounter it.

Tidbits—Fascinating little-known facts are highlighted in sidebars.

Activities—The "Sparky says" activities will help you to discover on your own a variety of new and interesting aspects of the North Woods and the wildlife that inhabit it.

Checklists—Species checklists are included at the end of each section.

Worksheets—Observation worksheets are located at the back of the book. You're welcome to reproduce these worksheets for use on your trips.

Style—In line with the informal style of this guide, common names are used in the headings and in the text. Those common names referring to a single specific species are capitalized, while generic common names are left in lower case. Since common names vary regionally, the scientific Latin name for each species is also included.

While all facts are based on the latest scientific research, the text has not been cluttered with references. Full citations for specific studies and key books are listed for your convenience in the appendix.

Wildlife watching tips

Watching wildlife in the wilderness is not easy. When we enter the wilderness, we are the strangers, and more often than not, the wildlife end up watching us.

Great Gray Owls can hear a vole skittering through its ice tunnels under three feet of snow. Timber Wolves can smell a Moose 300 yards upwind. Bald Eagles can spot a shallow-swimming fish one mile distant. With creatures of such ability watching us, it's no wonder we don't encounter more animals on our canoe trips.

You can, however, learn skills to increase the odds of your wildlife-watching success. Here are a few tips for making yourself a gracious guest in the wildlife's home territory and for seeing more of what's around you when you're out there.

Gather information about the lifestyle of the critters you expect to encounter. When are they active? What do they eat? Where are they found? This book will give you a good start.

Blend in, be quiet and sit still. Wear clothing that helps you blend into the surroundings. Stalk the animal by staying downwind and moving slowly and silently.

Follow the signs. Look for tracks. Study the markings and interpret the story they tell. Start where it's easy—on the shoreline or sand. After you've practiced, look for tracks in the woods where they're harder to discover. Examine the droppings of insects, birds, and mammals for evidence of their presence and their eating habits. Look for patches of fur, leftovers from meals, nibble marks on bark—even skeletons of the deceased—to discover the animals' homes. These are clues to how the animals live.

Watch the clock and the seasons. Like humans, animals develop a predictable rhythm to their days—regular times for eating, singing, exercising, playing, and resting. The seasonal cycles also occur regularly and predictably from year to year. Pay attention to the timing of what you observe. Then set your wildlife watching schedule in line with what you have learned.

Attract the wildlife you want to observe. Most animals are just as curious about you as you are about them. So look for ways to attract them. Clench your teeth and repeat the word "pishhh" four or five times. This sound drives some birds crazy with curiosity. Make a high-pitched squeak by kissing a knuckle on the back of your hand. This sound resembles the cries of an animal in trouble and will attract those in search of a free meal. Mimic the sounds that you hear—songs of the birds, calls of the animals. Bait your observation area with smells and food that will attract specific species.

Enhance your senses with the aid of technology. Use a magnifying glass to make the miniature world larger. Use binoculars to bring distant events closer. Turn on a flashlight or headlamp to help yourself see in the dark. Strap on a snorkel and mask to discover the underwater world.

So go ahead and develop your own style of seeing—and of being—in the woods. Make yourself at home in the wilderness. Sharpen your skills of observation and your senses.

A final note

Although specifically developed for the Boundary Waters and Quetico lands, *Canoe Country Wildlife* describes the most common creatures of the North Woods—many of whose ranges span the continent. This book will, therefore, be useful from Alaska and British Columbia to Quebec and New England, as well as around the upper Great Lakes. Bring it with you on journeys to Yosemite, Yellowstone, Banff, Algonquin Park, the Allagash, or the Adirondack Mountains—as well as on your excursions through Northern Wisconsin, Upper Michigan, and the Boundary Waters and Quetico wilderness.

Read and enjoy the essays. Try out the suggested activities—and experiment with some of your own. Use the checklists to record your observations of the wildlife you encounter. Consult the seasonal calendar to check out the natural events you're likely to witness. Then share with others what you learn through your adventures.

Canoe Country Wildlife is designed to be used. Cram it into your trusty Duluth Pack or stash it in your backpack, but wherever you carry it, keep it handy and use it! I hope it will make your North Country trips more enjoyable.

Mammals

Bats
Order Chiroptera

Flying mammals

Color varies: brown, black, gray, reddish

Often seen at dusk about lake margins

Unlike flying squirrels, which can only glide, bats can truly fly. A bat's "wings," actually skin membranes, stretch the length of the arm between four very long fingers and are attached to the sides of the body and hind legs. The thumbs are free, and hang from them. Bats can see but rely heavily on sonar for navigation and location of prey. Their prey is chiefly flying insects—moths, beetles, flies, true bugs, and mosquitoes. Bats, of course, are nocturnal. People have many false beliefs about bats. Contrary to their bad reputations, bats do not get tangled in people's hair, nor are all bats rabid.

The canoe country's seven species of bats all stay around during the summer, but when the bugs go, so do the bats. Silver-haired, Red, and Hoary Bats, known as "tree bats," migrate south to more "buggy" climes. "Cave bats" wait out the winter in the comfortably numb state of hibernation. These include the Northern Myotis, Little Brown and Big Brown Bats, and Eastern Pipistrelle.

Echo . . . echo . . . echo . . . echo

Bats are not blind—they do have eyes and can see well during daylight, but they must lock into sonar for night navigation and hunting. Cruising its favorite feeding haunts, a bat sends out ten to twenty high-pitched calls every second. The high-frequency sounds (up to 98,000 cycles per second) bounce off objects and return to the bat as echoes. This process is known as echolocation. When it receives a blip on its mental screen, the bat increases the sonar rate to 250 calls per second in order to fine-tune the picture. As the details become clear, a bat can even distinguish a flying beetle from a soft-bodied moth.

Snatching up to three hundred insects an hour, a bat can eat half its weight in insects a night. Contrary to the popular image, not all prey is caught by the bat's gaping mouth. A bat scoops most insects into its ballooned tail membrane and then transfers them to its mouth while on the wing. A bat can repeat this awkward performance twice in one second.

	Little Brown Bat	Big Brown Bat	Northern Myotis	Silver-Haired Bat	Red Bat	Hoary Bat
Wingspan & Flight Characteristics	8-10 inches low zig-zag flight	12-14 inches strong, steady flight	10-12 inches	10-12 1/2 inches slow and low	11-13 inches long pointy wings	14-16 inches long narrow wings
Color	uniformly glossy brown	uniformly glossy brown	dark dull brown	long pelage, frosted with silvery white	bright reddish/ orange to chestnut	brown with silver tips
Favorite Food	moths, beetles, mosquitoes	true bugs and beetles	variety of insects	insects hatching from streams	moths, beetles, and flies	moths, also beetles, bugs, and flies
Hunting Times	late dusk	active all night	dusk and dawn	early in evening	early in evening	active all night
Hunting Pattern	usually over water	over water and wooded clearings	over trees and ponds	zig-zag flight over streams	near streams and woodlots	wooded areas
Daytime Roost	caves, buildings, and hollow trees	caves, buildings, and hollow trees	caves, buildings, and hollow trees	roosts singly in hollow trees	roosts in trees, hangs by one foot	roosts singly in tall woody vegetation
Solitary or Communal	maternity colonies of several thousand	maternity colonies	maternity colonies of up to 30	solitary	solitary	solitary, except in migration
Migrate/Hibernate	hibernates	hibernates	hibernates	migrates	migrates	migrates
Winter Home	caves and mines Sept-May	buildings or caves where temps are just above freezing	migrates south to hibernate in small groups	winters as far south as northeast Mexico	moves south in loose groups	believed to winter in Mexico, sexes migrate separately
Number of Young	1 young (rarely 2)	2 young born June/early July	1 young	2 young born June/early July	1-5, usually 3-4	2 young in late May/early July
Potpourri	One lived just over 30 years, a record for bats.	Audible chattering is low-pitched radar signals.	Found in caves on Lake Superior's North Shore.	Returned home after being released 94 miles away!	Summers from Canada to South America.	Most adults in the canoe country are females.

Sparky says: Humid, still summer nights with a full moon provide the best times to observe bats. Sit on the rocks or paddle the shore quietly just after sunset. The mosquitoes may be swarming, but don't curse them; remember that they are bat chow. Watch the bats. How do they fly? Direct? Meandering? Zigzagging? How high above the water do they fly? How wide is their wingspan? Use the bat chart to make an educated guess as to which kind of bat you're watching. Even experts need a bat in the hand to accurately determine its species, so relax and have fun. Bring a big flashlight to spotlight the action.

Bat guano and the U.S. Army

Bat guano (poop) is harvested (mined?) from Texas caves for use as a nitrogen-rich fertilizer. The U.S. Army also uses guano. Guanine, extracted from guano, is combined with sulfuric acid to form nitroguanidine, which is used as a propellant in artillery and tank ammunition.

Black Bear

Ursus americanus

Black fur or cinnamon, blue, and brown in rare instances

Adults weigh 150 to 400 pounds

Often seen around campsights

Occasionally seen swimming

As a Boundary Waters–Quetico paddler the first Black Bear you're likely to encounter will be at the campsite. And that bear will not be there to see you. It will be much more interested in your trusty Duluth Pack and the powdered spaghetti sauce, Jell-O cheesecake mix, and peanut-M&M-raisin gorp inside.

Your first face-to-face meeting with a 150-to-400 pound Black Bear is likely to be intimidating. Razor-sharp claws and bone-crushing jaws that possess some mighty nasty looking flesh-piercing fangs seem to indicate that raw human might be on the menu ("mmmm, chewy on the outside, crunchy in the center"). But the mainly vegetarian Black Bears thrive on fruit, nuts, and grass. Eating grubs, ants, bees, snakes, mice, and nestling birds hardly qualifies them as vicious carnivores.

An average canoe country bear weighs between 250 and 325 pounds, is $4^1/_2$ to 5 feet long, and stands 2 to 3 feet tall at the shoulder. A 583-pound bruin established the Minnesota record, while a Wisconsin hunter bagged an 800-pound monster. The standard model is black with a white throat blotch, but Black Bears also come in blue, brown, and cinnamon phases. Grizzly Bears are not found anywhere east of the Rocky Mountains.

An em-"bear"-assing tale

This em-"bear"-assing tale is almost un-"bear"-able to recount. It happened eight years ago, and I can "bear"-ly remember it. I, an assistant guide, and seven hardcore juvenile delinquents from Chicago camped on a lake along the Granite River route. My fellow guide and I sat up late discussing tomorrow's itinerary, and suddenly we heard the telltale snap of twigs, which in the canoe country can mean only one thing . . . bear! We gathered up stones, flipped a canoe on its side to act as a "bear"-icade, and watched the show.

The other guide had insisted that I was hanging the food pack too low, but I had countered that there was no way a bear could reach it. This one, however, was a big Black Bear. Standing on his hind legs, he latched his claws into the faded green canvas and yanked it clean from the tree as if snatching a sock off a clothesline. We took bets on what would be eaten. We both lost. That bear ate everything, including our shortening. Truly an undiscriminating palate.

Take the proper precautions by hanging your food pack high from a limb or suspended between two trees. Keep the pack at least 10 feet up and away from the trunk. A lazy option is to put your pack under an overturned canoe with a stack of cooking pots on top. Leave no food in your tent. To drive off an unwanted furry campsite visitor, yell, shout, bang pots, and throw rocks. But if this doesn't work—and it often won't if food is accessible—just let the bruin go about his or her business. No use getting yourself mauled over some chunky peanut butter.

The big snooze

I have seen hundreds of bears in my brief lifetime. Unfortunately, 99 percent of them have been feeding at dumps or engaging in other human-assisted food-gathering operations (i.e. foraging in garbage cans or at bird feeders, or taking roadside handouts).

One encounter with a bear doing bear things stands clear in my memory. One Halloween my friends and I were paddling down the Little Indian Sioux River just south of Devil's Cascade. I looked across the wide cold river and saw a Black Bear drinking from the shore. "Bear!" I yelled. The sow angled up the hillside over the golden aspen leaves. We followed. Under the upturned roots of a fallen Quaking Aspen we found the fresh diggings of a bear den. From this evidence we assumed we had seen a female. Sows need the comfort of a leaf-lined, well-protected den to give birth.

Male Black Bears tend to hibernate where they fall, so to speak, with minimal shelter. The base of a tree serves as well as a pile of brush. I've talked to people who have snowshoed right over the top of a hibernating male bear.

The controversy over whether Black Bears are true hibernators or not seems to be inconsequential. While their body temperature and pulse rate drop only slightly, bears do not drink, feed, urinate, or defecate during their six-month slumber. Now that's a deep sleep. A half year without food requires bears to put on a thick blanket of fat in the fall. They travel to exploit abundance. They seek carbohydrate-rich blueberries, chokecherries, Wild Sarsaparilla berries, hazelnuts, and acorns in order to accumulate the 100-pound layer of fat needed for a comfy cozy winter sleep.

Most bears are asleep by mid-November. They cut their metabolic rate by half yet maintain an 88° F body temperature, which is within 12° of normal. Females awake in January just long enough to give birth to two or three tiny 8-ounce cubs. The equivalent would be human mothers birthing 4-ounce babies! The sow bear, usually, falls back to sleep. The tiny cubs nurse and grow but do not hibernate.

It is truly amazing that bears can survive a half year without voiding body wastes. Urea, a substance that would quickly poison the human body, breaks down in the sleeping bear to create protein that helps maintain muscle tissue. Bears actually awaken with more muscle than when they went to sleep. Now there's a great idea for a health spa. Join up, come in to sleep for a couple months, and leave with a well-toned body! Physiologists have taken a keen interest in this phenomenon, since the voiding of wastes poses one of the big problems with cryonics, long-term human "hibernation" by freezing.

Bear biology is crucial to current cholesterol research. The Black Bear's cholesterol rate is two times that of a normal human but they suffer no hardening of the arteries or gallstones. Researchers found that bears produce a bile juice that dissolves gallstones in human test patients.

Mid-April finds most of the canoe country's bears awake and stretching in the spring air. Some will have lost a full 30

Eats like a bird

Contrary to popular belief, Black Bears are not bloodthirsty carnivores, killing other mammals to satisfy their hunger. Rather they subsist on fruits, berries, insects, grass, and flowers. Opportunistic to the end, though, a bear will not pass up a nestful of young birds, an old deer carcass, or a snake slithering across its path.

percent of their body weight—their 100-pound fat reserves. Mothers and their cubs hang out at the bases of large White Pines, whose ridge-barked trunks offer an easy climb to safety for tiny cubs. Food is scarce until the green-up, which begins about May 1. Aspen and willow catkins provide early spring appetizers for the bears, holding them over until they can graze on lush spring grass. The fattening process starts all over again.

 Sparky says: Pick berries and make a wilderness pie. Enjoy the canoe country's brief abundance of lush, plump blueberries, raspberries, juneberries, and chokecherries, just as the bears do each summer. Wild blueberries put pulpy domestic varieties to shame, and a pie of your own hand-picked berries is a special treat. Look for charred standing snags that signal recent burns, for here raspberry and blueberry plants thrive. The Fourth of July marks the time to start looking for early ripe blueberries, but they will still be "pickable" through early August. Fill a couple pots or empty water bottles with berries to take back to your campsite. Relish them in biscuits and pancakes, a cobbler, or a pie. Here's an easy camp recipe for blueberry bannock:

Mix	2	cups	flour
	3	tsp.	baking powder
	1/2	tsp.	salt
	6	tbsp.	shortening/margarine
	4	tbsp.	powdered milk
Add	1	cup	blueberries
	1/3	cup	water

Shape into 1-inch-thick cakes. Dust with flour before placing in a hot, greased fry pan. Turn when bottom is golden and crusty. Check cakes for doneness by inserting a wood splinter. If it comes out clean, they're ready to be smothered in butter and slowly savored.

River Otter

Lutra canadensis

Large aquatic weasel

Up to 4 feet long from nose to tail

Often seen in groups, only their heads sticking out of the water

Snorting, gurgling, and churring noises

The River Otter is the canoe country's second-largest member of the weasel family, smaller only than the rare and much maligned Wolverine. It may surprise some to learn that not only are River Otters and Wolverines weasels but so also are Striped Skunks, Badgers, Mink, Fishers, and Pine Martens.

How large are River Otters? A big Alaskan male may be 5 feet long from sniffing nose to twitching tail and top 30 pounds. Most do not exceed 4 feet and 20 pounds. Their true size comes as a shock to many paddlers, who usually encounter otters as just curious heads, sniffing, snorting, grunting, purring, and growling from the water. What you see is just the tip of the iceberg—or literally, of the otter.

Campers commonly see River Otter families swimming in lakes during the summer. Their pelage is lustrous brown with a silvery wash on the muzzle and throat.

Slip sliding away

River Otters are first-class boreal buffoons. Much has been written about whether or not animals play, and some behaviorists remain skeptical, but it would be hard to watch a group of otters repeatedly sliding down a snowy bank on their bellies and not believe that they were having a darn good time.

The otter's method of winter locomotion incorporates "the slide," since simple bounding is too boring and slow. An otter runs only long enough to get up speed to fling its torpedo-shaped body into a graceful glissade. It lopes-and-slides, lopes-and-slides, lopes-and-slides, in snow and on ice. One naturalist measured the slide of an otter on light snow over smooth ice at 27 feet!

It always amazes me to see their tracks so far from any apparent open water. Otters may keep ice-holes open in frozen lakes through repeated use. Gaps of air between the bottom of the lake ice and the lowering lake level allow otters to surface for air beneath thick ice.

Mustelid chow

River Otters are aquatic, as are most of their prey. Crayfish and fish compose the bulk of their diet. An undulating body propels the otter down to the lake bottom, to a maximum depth of 45 feet, where it noses its way among the rocks, with long, sensitive whiskers, feeling for crayfish, hibernating frogs and turtles, or a torpid bass. An otter can stay submerged for two minutes.

Muskrats beware! River Otters are known to enter Muskrat houses to kill and eat the occupants. Now that's an eviction!

Intestinal fortitude

The intestines of River Otters have a thick mucous lining that prevents puncture by fish bones.

Sparky says: Look for telltale signs of otters along the canoe country's lake and river shores. They often deposit their droppings on rocks next to the water's edge. Fresh droppings are wet, black, and mucous-covered, becoming white and powdery when dry. Try dissecting a dropping to find out what an otter had been eating. Look for crayfish claws, mammal hair, fish scales, and bones.

Pine Marten

Martes americana

Length 1 to 1 1/2 feet plus a
7-to-10-inch tail

Chocolate-brown weasel

The Pine Marten is undoubtedly the cutest animal in all the North Woods and unquestionably the most adorable weasel in all the world. Its kittylike face, lustrous chocolate-brown fur, graceful bounding, and inquisitive nature endear it to many people. (To their frequent victims, the Red-backed Vole and Red Squirrel, however, they are probably the most hideous of creatures.)

But I'm biased. Several generations of martens were raised literally under my feet, beneath the floorboards of my cabin. I heard the kits growling and wrestling day in and day out. Their mother led the way to the bird feeders, which we regularly stocked with scrap chicken bones. The young were at ease with us even to the point of napping on the bird feeder in broad daylight.

Smaller than cousin Fisher but larger than cousin Short-tailed Weasel, the Pine Marten has a 1-to- 1 1/2-foot-long body attached to a 7- to10-inch tail. Unlike their aquatic cousins the Mink, the hydrophobic Pine Martens loathe water. Swimming mats their unoiled fur, and they'll go to any lengths to avoid getting wet.

Though usually chocolate-brown with a golden throat patch, marten can appear nearly black and, on rare occasions, completely blond.

Old-growth-forest dwellers, the Pine Marten population declined precipitously after intense logging that took place around the turn of the century. Their numbers had already been depleted by trapping in the 1700s and 1800s for exportation to fur-hungry Europe. In 1856, a fur trader near Lake Superior purchased one thousand marten pelts. By 1933 the marten was so scarce it required total protection in Minnesota. Due to maturing forests, the Pine Marten has made a dramatic comeback since 1970, to the point that a limited trapping season has been allowed since 1985.

Fortunately, much of the canoe country was spared from logging due to the rugged inaccessibility of the land. This provided a haven for the marten. The canoeist may encounter marten on portage paths or near campsites in mature spruce-fir forests. If you encounter one, give a quick "pshhhh" or squeak on your knuckle to stop it in its tracks. Anything that remotely sounds like a critter in distress will pique the curiosity of this inquisitive hunter.

Arboreal acrobatics and treetop tricks

Wabachis, the rabbit chaser, is the Cree name for the marten. But Snowshoe Hares make up only a small portion of the Pine Marten's diet. Red-backed Voles probably provide the main course for Minnesota martens.

Sticking its nose into every crevice and under every log, the Pine Marten searches randomly through the deep woods. In winter, slightly offset paired tracks in the snow confirm this. Using fallen logs as runways, Pine Martens dip in and out of snow burrows, through brush heaps, down logs, and into root tangles. When the mouse-vole population cycle is at its peak (many mice in the woods), marten territories may be as small as 1/4 square mile. When mice become scarce, a marten may need to roam a huge area, up to 10 square miles, to find sufficient food. Males and females occupy separate territories, but when territories overlap males hunt in the evening and females during the day.

The opportunistic Pine Marten will eat most any small critter it encounters in its travels. Voles, mice, shrews, chipmunks, flying squirrels, birds, and insects are all acceptable food items. Surprisingly, raspberries, when in season, may constitute up to 85 percent of a marten's diet.

Pursuit and capture of treetop Red Squirrels is the Pine Marten's special, though rarely utilized, talent. The marten is the only mammal that can catch a Red Squirrel in the tree, matching its arboreal acrobatics leap for leap. Hind legs that rotate backwards, just like a squirrel's, allow the marten to scamper headfirst down trees. Even with this distinct ability, a marten rarely takes Red Squirrels in the trees, preferring to nab the critters in winter when they are on the ground digging for their seed caches. Voles are much easier to catch, so Red Squirrels comprise only 3 percent of marten diets.

Delayed implantation

Male Pine Martens prefer the single life eleven months a year. They associate with the females only during the brief but intense midsummer mating season. The lively courtship consists of play and wrestling, punctuated with growls, screams, and "love chuckles," all consummating in copulation. This summer rendezvous results in three or four naked, deaf, and blind kits being born in March or April, eight to nine months later. This is an extremely long gestation period for a mammal this size. Researchers explain that the fertilized eggs are not implanted in the uterine wall until several months after initial mating. Only during the last month does a spurt of growth take place in the embryos. The phenomenon, known as delayed implantation, is a characteristic of all weasels, many rodents, some bats, one deer species, Black Bears, and Nine-banded Armadillos.

Sparky says: "Fox walk" your way to discovery. Tracker Tom Brown coined this term for a silent walking technique that he learned from a Native American mentor. Keeping all your weight on your rear foot, step slowly forward with the other, gently and silently allowing the front foot, ball-first, to come in contact with the ground. Before placing any pressure on the foot, make sure no twigs are underfoot. Slowly transfer all your weight to that foot and repeat. One step may take five minutes. It is an excellent technique for stalking wildlife. Remember, move only when your quarry looks down or away.

My buddy and I watched a Mink hunt on the shoreline of Cross Bay Lake. It appeared from behind a fallen log just an arm's length from us. We watched it swim along a granite cliff. The Mink investigated every crevice in the rock. Suddenly it dove, and air bubbles rose to the surface. It resurfaced with a crayfish, which it promptly ate in the shelter of a rock. All we could hear was the crunching of the exoskeleton. We later found that the Mink had consumed the entire crayfish except for a few of the hind legs.

The Mink's next hunting ground was a shoreline rock pile. I made a squeaking sound by sucking against my knuckle. It worked. Fooled into believing I was an injured bird, the Mink promptly appeared, its brown fur soaked and matted and gave me a quizzical look. Satisfied that I was not a potential victim, the Mink raced off, long tail trailing, to try its paw at the terrestrial hunt. A chattering Red Squirrel marked its location further down the shore.

Mink are weasels. Long, slender, and sinewy, Mink reach 16 to 29 inches tip to tail, the tail accounting for a third of its length. Males are 10 percent larger than females. They are dark brown year round, with a white blotch on the throat.

Members of the weasel family abound in the North Woods. Identification can sometimes be confusing. Otters are much larger, communal, and more aquatic. The Pine Marten is solitary, like the Mink, but strictly a landlubber. The Long-tailed Weasel, Short-tailed Weasel, and Least Weasel are much smaller and more terrestrial.

The canoe country provides Mink with thousands of miles of fascinating shoreline for hunting. Watch for them during daylight hours of the ice-free months. Sit quietly on a rock with a lake view and wait. Quiet is the key.

Loners

Mink are real loners. Males set out on solo hunting trips that may take them seven days to complete. Females also make

Mink
Mustela vison

Dark brown fur

Length 16 to 29 inches

Rarely seen away from shorelines

At home on the ranch

Say it plain: It takes seventy-five adult Mink pelts to make one full-length fur coat. Though wild North Woods Mink are widely trapped, it is mainly their brothers and sisters down on the Mink ranches that become the fur coats of tomorrow.

foraging loops, but they stay closer to home. Males and females get together only to mate.

Their quarry varies: frogs, crayfish, birds, salamanders, voles, mice, snakes, fish, ducklings, eggs, young Muskrats, young hares, turtle eggs, and an occasional Red Squirrel. Efficient killers, Mink twine about a larger victim in order to finish it off with a fatal bite to the neck.

Sparky says: Sit down on your favorite piece of shoreline with a pair of binoculars, open eyes, and an alert mind. Sit quietly for at least one hour. Watch the world. Try it at dawn, dusk, and noon. When did you see the most? If you are fortunate, you too may encounter the far-ranging Mink.

Although rarely seen and rarely heard, *Canis lupus* survives and thrives in the canoe country wilds. Currently considered federally endangered in the United States, the Timber Wolf is merely "threatened" in Minnesota. Fifteen to eighteen hundred wolves roam the northeast part of the state, the highest population in the lower forty-eight states. They are so wary and the packs so widely spaced that summer Boundary Waters–Quetico visitors rarely encounter wolves face-to-face. However, hearing a pack howl, an experience possible any night of the year, can be as spine-tingling as seeing one.

Be on the lookout for wolf sign: large droppings consisting of bones and hair, tracks five inches long or longer, scent posts of scraped earth and urine, and partially eaten deer and Moose carcasses. To increase your chances of encountering wolves in the canoe country, bring a quiet spirit, watchful eyes, and a big dose of good luck.

The Eastern Timber Wolf subspecies is not as large as the Alaskan but is far bigger than the extirpated Mexican subspecies. Adult males average 70 to 85 pounds but may often top 100. The record wolf weighed in at a hefty 210 pounds. Females average 15 pounds less than males and rarely reach 100 pounds. Standing 26 to 32 inches at the shoulder, wolves carry their bulk on long, slender legs. They have an interesting, casual gait that from behind looks almost floppy. Most Boundary Waters–Quetico wolves are grizzled gray; black wolves are occasionally seen, but white ones are few and far between.

How do you tell a North Woods Coyote from a Timber Wolf? It's not easy but here are some general rules. Coyotes have shorter legs, longer ears, a pointier nose, and a smaller body than do wolves. Wolves lack the red and yellowish tones that Coyote fur possesses. Timber Wolf tracks span 5 inches, twice the size of a Coyote track. Howling wolves use long, drawn-out single tones while Coyotes yip, yap, and yowl.

Timber Wolf

Canis lupus

Height 26 to 32 inches at the shoulder

Size of a large dog

Grizzled gray

Howls with long, simple tones

Lives in extended family groups called packs

Wandering Wolf

Number 1239 was a young adult Timber Wolf when he was ear-tagged and radio-collared by the United States Fish and Wildlife Service in July 1980, just south of International Falls, Minnesota. Fifteen months later, 1239 was dead, shot by a hunter near Carrot River, Saskatchewan, 551 straight-line miles north-west. Number 1239, a restless soul, in the process of his wanderings set a new record for wolf travel. The journey will also go down in scientific journals as one of the longest documented trips any land mammal has ever made.

A family affair

One of North America's most family-oriented animals, wolves live in a pack, a complex social unit that operates on a dominance hierarchy within the structure of an extended family. The leaders of the pack, the alpha male and the alpha female, are dominant over all other pack members and are the only breeders. Hierarchy equals harmony in this family. When all know their status within the pack, infighting is curtailed. For example, a hypothetical, hierarchical, harmonious pack has just killed a Moose. Alpha male and alpha female feed first. The beta male gorges next, followed by the young subdominant adults. The yearlings (last year's pups), also known as "adolescent rowdies," have their own little social stratum led by the little alpha. They pick over what's left, leaving the scraps for the "scapegoat," an extremely submissive male or female wolf. The adults feed any pups present.

Wolves display and maintain status by body posture, facial expression, vocalization, and, most important, tail position. The higher the tail, the higher the wolf's status. The ultimate sign of submission is to hold the tail down between the legs while rolling over onto the back. (Dogs use the same social signals.) But the hierarchy, far from being static, is in the process of constant change. The beta male could challenge and defeat the alpha male at any time, instantly becoming the new leader of the pack. Battles for dominance start early. Tiny three-week-old pups wrestle for position among their siblings. As juveniles mature, they must choose to be "biders" or "dispersers." Biders stay with the pack in a submissive role, biding their time until an opportunity comes to move up. Dispersers leave the pack to become lone wolves. They may join another pack or find a mate and start their own pack.

Packs, which in the canoe country average seven or eight members, mark and defend an average territory of 48 to 120 square miles. Borders are marked with scent posts, areas of scratched earth near a conspicuous rock or tree where the wolves have urinated. Neighboring packs see and smell these posts as traffic signals. Fresh scent is a red light that means "this area closed." An old scent post signals green, "go ahead." Anything in between is treated as a yellow light, meaning "proceed with caution." Lone wolves skirt the edges of the territories, lest they be killed as an intruder. The buffer zones between territories are also traveled by Coyotes and are safe havens for prey animals. Territory boundaries are flexible and change as packs change. Prey concentrations of deer and Moose also affect boundary size. When prey is scarce, wolf packs expand their territory to include more prey animals.

Ayooooooooo!

Ethologist Fred Harrington, after years of research on wolf behavior, never could make any connection between wolf howls and the phases of the moon. But he did find definite seasonal howling peaks. March is one, the mating month when testosterone levels run high. Howling activity peaks again during August and September, when the pups are three months old and vulnerable.

Howling serves the pack as a long-distance early warning system. According to Harrington a howling pack says to its neighbors, "We are a pack, we are here, and we intend to stay here. If you get any closer we might attack." It serves to prevent any unplanned encounters and hence, unnecessary conflicts. The message applies only to other wolves, though Moose, White-tailed Deer, Beavers, and Snowshoe Hares certainly listen very carefully.

A howl starts low and rises in pitch as the wolf raises its muzzle skyward. If two wolves hit the same note, one immediately changes pitch. Five wolves howling on completely different notes sound like a much larger pack than do five wolves on the same note. This bluff can come in handy when dealing with unruly neighbors.

Meat on the Hoof

Don't pity a healthy Moose surrounded by a pack of wolves. One well-placed kick with a sharp hoof can crush a wolf's skull. Wolves, though, have the uncanny knack of singling out sick, old, and weakened animals that to you or me would look perfectly healthy. If the prey is brought down, Timber Wolves choke down 18 to 20 pounds of meat (nearly a quarter of their own weight), then take a long nap (15 to 20 hours). Wolves even crack open bones to get at the edible marrow. The pack won't hunt again for two to four days. The carcass also provides a feast to the boreal cleanup crew: Coyotes, Red Foxes, Pine Martens, Blue Jays, Gray Jays, weasels, ravens, crows, and chickadees.

Sparky says: Try howling on a lonesome lake during the peak howling months of August or September. Howling can be fun for the whole family. But you have to take it seriously. Wait until nightfall on a still evening. Face the immense black quiet. No laughing or giggling allowed. Start low, and go higher, and sustain. Don't yip like coyotes. Everyone should be on a different pitch. Ten to fifteen seconds should be enough. Quiet. Listen. Listen. Stand perfectly still and listen to the night. Listen for a good minute then try again. Don't expect an answer, but hope for one.

Red Squirrels take life seriously. They always seem to be chattering at someone or something. Some call them "chatter bears." They get very touchy about intruders in their feeding territory, and if a good scolding doesn't do the job a spectacular chase ensues. The pursuit is silent and fluid, and often crosses the path of canoe country visitors.

"Oh, look at the baby squirrel!" exclaim first time visitors from the deciduous south, where the huge, and equally red, Fox Squirrel lives. But full grown at 8 inches long, with an additional 5 inches of tail, the Red Squirrel is less than half the size of the familiar Gray and Fox Squirrels. Red Squirrels' pelage changes with the seasons. In winter they wear bright red-orange, with prominent ear tufts. They lose their bright color and eartufts in summer, but a black strip then separates the white belly from the dull red back and sides.

Hoarding

Red Squirrels are winter-active. They require a reliable supply of power food to fuel their nervous bodies through the cold, dark months. Cached conifer seeds are the answer.

The "thunk . . . thunk . . . thunk" of falling White Spruce and Jack Pine cones and branchlets is an annual North Woods event that begins the first week in August. I timed one ambitious Red Squirrel, who dropped individual spruce cones at a rate of one cone every $1^{1}/_{2}$ seconds. Occasionally they nip off cone-bearing branchlets. I saw fifty to seventy such Jack Pine branchlets, measuring 3 to 8 inches long and bearing unripened cones, under one pine. Maybe the squirrel finds it easier to bite off entire branchlets, rather than individual cones while perched far out on a limb, or possibly the attached greenery makes the cones easier to find once on the ground. Either way, the cones and their edible seeds are gathered and stored in caches within the squirrel's territory. Mushrooms, even ones poisonous to humans, are also cached and hoarded. In summer you may have noticed a mushroom wedged into the low branches of a

Red Squirrel
Tamiasciurus hudsonicus

Small squirrel with rust-colored fur

Length 8 inches plus a 5-inch tail

Chatters at intruders from low perches

Cradle robber

Not normally thought of as carnivorous, a Red Squirrel may eat up to two hundred bird eggs and nestlings per year. But that's the nature of nature.

Sap suckers

Red Squirrels crave tree sap in the spring. Oozing sap from holes drilled in Paper Birch trees by Yellow-bellied Sapsuckers is quickly lapped up. They also gnaw through the bark of Sugar Maples to get at the sweet sap.

bush or tree. A Red Squirrel put it there to dry before being stored. This activity makes squirrels important dispersers of fungal spores.

A typical territory may contain fourteen thousand stashed cones and mushrooms, enabling the well-prepared squirrel to survive two years of cone crop failure. Storing cones in hollow logs and stumps prevents them from drying and prematurely losing their seeds. Squirrels locate larders under the snow by memory and smell. Researchers believe that Red Squirrels can smell cached cones through 12 inches of snow. Fierce defense prevents others from raiding the stores. The gathering, storage, and defense of these precious caches allows the Red Squirrel to survive the boreal winter.

Sparky says: Go on a midden hunt! What is a midden? It is the Red Squirrel's garbage dump, made up of the discarded cone scales. The squirrel sits atop the pile and gnaws a cone apart to get at the seeds inside. A large, often-used midden may be composed of a bushel of cone scales. Look for middens atop stumps and moss hummocks and on downed logs.

Northern Flying Squirrel
Glaucomys sabrinus

Nocturnal squirrel

Mouse-gray pelage

Large eyes reflect red-orange

The Northern Flying Squirrel is probably the canoe country's most common squirrel, a fact that surprises day-active humans. But if we were as nocturnal as flying squirrels, we would fully comprehend their abundance. The rare encounters we have with them in the wild are often limited to our hearing them: a scampering up a rough-barked pine, spruce, or fir, then a pause for the glide, and a little "whoomp" when the squirrel lands on another trunk. Walking down a trail one night, my friend was mistaken for a tree trunk by a flying squirrel who landed smack on his chest. Fortunately, they only weigh about as much as a heavy whole-wheat pancake.

I've watched many "fairy diddles," as they are called in some parts, at the sunflower-seed feeders on Seagull Lake. They glide in from the dark forest to land vertically on the lodge before scampering to the feeders. The outdoor lights are on, but the squirrels remain unfazed. They are beautifully delicate creatures. Improbably large brown eyes, for light gathering in dim conditions, stand out from their mouse gray and cinnamon brown velourlike pelage. In other words, they are darn cute. And tame too. On several occasions I've come close to hand-feeding them. They treasure sunflower seeds.

Your best chance to view a flying squirrel while on a canoe trip is to carry a flashlight on a night hike to the latrine. If you hear the sounds of a critter scampering up a tree, shine the light on it. Flying squirrels' eyes reflect red-orange. A fortunate person may actually see the flying squirrel glide away.

To fly

The Northern Flying Squirrel may be more appropriately called the northern "gliding" squirrel. Its flight is actually a descending glide on flaps of skin connecting the wrist and ankle. These flaps, called patagia, are spread wide in flight, effectively tripling the gliding surface area to create a personal parachute. Bats are the only other Northern American mammals to possess patagia.

Sweet and sorrow

In spring, flying squirrels are attracted to maple sap buckets. Just looking for a sweet drink, they occasionally fall in and drown. Flying squirrels are poor swimmers due to their loose flaps of patagial skin.

When it becomes too dark to see the bark of a tree from 6 feet away, it is dark enough for flying squirrels to emerge from their abandoned-woodpecker-hole homes and face the gathering darkness. Clinging to the trunk, head down, the squirrel bobs its head, triangulating with large, wide-set eyes on a nearby target tree trunk. Leap. Forelegs and hindlegs spread its patagial cape wide, like a miniature Superman. Descend and glide, the tail ruddering the body into position. Descend and glide. At the last second it pulls up, with the body brought vertical as the cape balloons to act as a brake. A quick scamper to the opposite side of the tree deters any pursuing owls. Up to the top it scurries and off it glides again.

Their average glide is 20 to 30 feet, but they can occasionally pull off 160-foot monster glides. Flying squirrels can only descend. "Muscle memory" allows them to rapidly travel frequently used routes through the dark woodlands.

Homelife

Gregarious and affectionate creatures, Northern Flying Squirrels embrace during mating and kiss their fellow denmates. In winter, they nest communally but remain active. One naturalist discovered fifty Northern Flying Squirrels denned together in a hollow tree. Evidently claustrophobia is not a flying squirrel malady. They eat fungi, lichens, fruits, seeds, insects, an occasional bird egg, and the buds and blossoms of aspen, alder, and pussy willow.

Sparky says: Try for a glimpse of a flying squirrel. Day or night: scratch on the bark of a tree with an old woodpecker hole in it. Watch the opening for a curious flying squirrel head to poke out. Its expression may say, "Who's that climbing up my tree?"

Chipmunks in the canoe country are real campsite clowns. They've learned to associate campsites with humans—and with food. Hanging your food pack will keep out more than bears.

Spotting chipmunks is not difficult, more likely than not they'll be zipping all over the campsite. Trying to distinguish between the two species that inhabit the North Woods is the challenging part. The Eastern Chipmunk is about 5 inches long, with a 3-to-4-inch tail. Black and white stripes run down the back, but end at a reddish rump. When the chipmunk runs, the tail is directed backwards.

The Least Chipmunk, by contrast, weighs only half as much as the Eastern and has a much more delicate-looking body that is only about 4 inches in length. The back stripes, much more crowded together, run all the way to the base of the tail, which is held straight up when the animal scurries about. The stripes extend onto the face and are well defined. These contrasting stripes allow the chipmunk to blend in beautifully with the sun-dappled, jumbled mass of the forest floor. Oh yes, Least Chipmunks have five upper cheek teeth, while Easterns possess only four. But you better find a chipmunk dentist to handle that one.

The name "chipmunk" is corrupted from the Ojibwa name for them, *chetamon*.

A chipmunk can munk chips!

How many chips can a chipmunk chip? Well, probably as many as you can feed it. But, of course, I do not condone feeding wild animals, even if they are as cute as chipmunks. Besides, Eastern and Least Chipmunks are master hoarders, very capable of caring for themselves.

In summer, they enjoy leaves, buds, mushrooms, berries, insects (especially grasshoppers), frogs, slugs, fruit, tree seeds, nuts, plant tubers, and handouts. I have seen chipmunks eating pin cherries while perched four feet up in the tree. They do

Eastern Chipmunk
Tamias striatus

Least Chipmunk
Tamias minimus

Eastern Chipmunk

Black and white stripes end at rump

Tail extends backwards

Least Chipmunk

Length 4 inches

Stripes extend from face to tail

Tail held straight up

Russian chipmunks

In Russia, chipmunk-stored nuts fetch a higher market price than human-harvested ones because chipmunks are well known for their quality control and would never hoard a second-rate nut!

Rodent mansions

Chipmunk burrows are the mansions of the rodent world. Many encompass 30 feet of tunnel connecting a storage room, bathroom, resting den, nesting chamber, front entrance, and back door.

most of their foraging in the hours just before and after high noon, probably to alleviate competition with the morning- and evening-feeding Red Squirrels.

To prepare for winter, chipmunks cache up to a gallon and a half of tree seeds, nuts, and plant tubers in a large pile in the sleeping chamber of their underground burrows. Cheek pouches allow them to carry sixty to seventy sunflower seeds to expedite this process of stockpiling. This food pile serves as their bed. They sleep on it, and whenever hunger pangs awaken them they enjoy a "midnight snack." Unlike many hibernators they do not put on a thick layer of fat in the fall but instead depend on their "edible bed" for nourishment.

Sparky says: Use the description from the text to identify the species of chipmunk you have at your campsite. Younger children can watch their tails as they scamper about. Eastern Chipmunks hold their tails straight back, while Least Chipmunks hold them straight up.

The Beaver is the second largest rodent in the world. (The largest is the South American Capybara—I knew you'd ask.) Adult Beavers weigh 35 to 65 pounds, but monsters may top 100 pounds. Of course, today's Beaver would be dwarfed by the Giant Beavers of the Eocene Epoch of 55 million years ago, which measured 7 feet from tip to tail and possessed incisors 11 inches long.

Beaver

Castor canadensis

Large flat-tailed rodent

Weight 35 to 65 pounds

Almost completely aquatic

Create a distinctive V-ripple when swimming

Beautifully created for life underwater, Beavers possess a wet suit of a dense mat of underfur and guard hairs that they keep dry by secreting an oily substance from their abdominal skin glands. Valves shut in the nose and ears, keeping water out. Swim goggles, actually a transparent membrane, close over the eyes to give superior underwater vision. Lips shut behind their large yellow incisors so they can carry or gnaw branches underwater. They propel themselves with webbed hind feet that spread to create 7-inch-wide paddles.

Beavers have a special air passage separate from the larynx and can tolerate high levels of carbon dioxide in their blood. These adaptations allow the Beaver to swim a half mile underwater and stay down for twenty to thirty minutes.

The hairless, wide, flat tail has four main uses. Can you name them? If you said to carry mud or turn into a chainsaw to cut trees down, you've been watching too many cartoons. The uses are:

1. Alarm bell—they slap it on the water to warn others of an intruder.
2. Rudder—for steering underwater.
3. Kickstand—a prop when standing upright and chewing on trees.
4. Refrigerator—for fat storage.

Keep still as the sun melts into the northwestern horizon and watch for the unmistakable V-ripple trailing a swimming Beaver. Very common in the canoe country, Beavers are most active at dawn and dusk on smaller lakes and back bays of bigger lakes. Dams, lodges, and Beaver-cut trees provide obvious clues to the presence of this aquatic rodent.

Furry gold

Believe it or not, Beavers were the catalyst for the exploration and settlement of Canada and much of the northern United States. Basque fishermen first recognized the value of Beaver pelts when, while sailing the eastern shore of the new continent in the early 1500s, they encountered Indians wearing Beaver robes. The older robes were most highly prized because after extended wear the long guard hairs were worn off, leaving only the dense lustrous underfur. The fishermen returned to Spain with their traded-for prizes and all of Europe took notice. The race for Beaver was on.

Pierre Esprit Radisson and Médart Chouart, Sieur de Groseilliers, after being rejected by their own Quebec governor, went to England and found support for a 1668 expedition to Hudson Bay in quest of the furry gold. This led to the formation of the Honourable Company of Adventurers Trading into Hudson's Bay, otherwise known as the Hudson Bay Company.

The Hudson Bay Company, unopposed, did a booming Beaver trade. Their posts rimmed the great bay at Churchill, York Factory, and Moose Factory. The northern Cree trapped in the winter when the fur was at its finest and then made spring pilgrimages to the Hudson Bay Company posts to trade the pelts for blankets, beads, and guns.

This British monopoly didn't last. In 1783, a Montreal-based partnership of Scottish, English, and American businessmen joined resources to form a large, powerful fur-trade company called the North West Company. Partners McTavish, McGillivray, Frobisher, Todd, and Mackenzie were aggressive, innovative men. Instead of having the Indians come to them, the North West Company went to the Ojibwa, Cree, and Assiniboine villages, where they built temporary posts for trade.

They employed French Canadian farm boys to paddle the huge 25-to-45-foot birch-bark freight canoes, laden with up to 4 tons of trade goods, from Montreal to Grand Portage and beyond. The canoes resupplied the hundred or so North West Company posts scattered about the Canadian west. The colorful voyageurs paddled, smoked, sang, packed, grunted, bragged, and cursed their way along the "voyageurs highway" of Rose, Gunflint, Saganaga, and Knife lakes, which today form the U.S.–Canada boundary and separate the Boundary Waters from the Quetico.

The Indian people did all the hunting and trapping. They brought Beaver, Red Fox, Pine Marten, River Otter, Black Bear, Mink, and Timber Wolf pelts to the North West Company clerk, who would convert them into the monetary unit of the North West Company—the "made-beaver." One large Beaver pelt equaled one made-beaver. Two small Beaver pelts equaled one made-beaver. It took twelve squirrel skins to equal one made-beaver. The clerk recorded all this in the ledger next to the Indian hunters' names. They could then trade for the useful items they desired. A four-point Whitney's wool blanket cost four made-beaver. A flintlock musket cost between twelve and twenty made-beaver. The Indians were not cheated with trinkets but got many useful items that made everyday life a little easier. Cooking in lightweight, durable tin kettles was a real treat compared to using birch-bark pots. Starting fires with flint and steel was far easier than the old method of rubbing two sticks together.

After ice-out the pelts were canoed from the scattered posts to Montreal and then shipped to London. The North West Company alone sent out one hundred thousand Beaver pelts per year during its heyday. The hair was turned into top hats for the men of society in Paris, Vienna, and Stockholm. The hatter plucked out the long guard hairs first, and then scraped all the underfur off the hide. The pile of hair was vibrated into a loose mat. Then the hatter boiled and shrank the mat to make a tight felt. Tiny barbs on each hair locked with other barbs to hold the entire felt together and make it strong. No other fur possessed this felting quality, which made Beaver the most sought-after fur in human history. After forming and drying

The Beaver-Fish

Devout Catholic French Canadian voyageurs enjoyed eating Beaver and putting the fatty tail in their stew, but religious beliefs kept them from the practice on Fridays. Word of the problem got back to the Pope. A man sent to study the Beaver reported back that this strange animal swam in the water all day and had scales on its tail. So by papal decree the Beaver was officially declared a fish, and the voyageurs could eat it any day of the week.

3 x Beaver

Many Stetson cowboy hats are still made of Beaver-fur felt. Look for the "3x" to "20x" inside, which denotes the amount of Beaver hair used in the hat. The higher the "x" value the higher the percentage of Beaver hair mixed in to make the felt. The National Park Service has recently switched to using Nutria fur for their ranger hats. Nutria are South American rodents which are fur-farmed in the southeast United States.

Long days

Only dim light filters down through winter pond ice to reach the Beaver below. The result is a breakdown of circadian rhythms, causing the Beavers to readjust to a twenty-six-hour to thirty-hour day with longer rest periods to conserve energy.

over a hat block, the Beaver had completed its journey from wild and free rodent to a high-fashion top hat. A single hat cost a middle-classed Londoner between three and six months' wages. Gentlemen even passed down the highly prized hats in their wills.

Mercury was used extensively in the process, poisoning the hatters and causing mental disorders. No one knew the cause of their illness back then but they became universally known as the "mad hatters."

In the early 1800s, silk replaced Beaver as the luxury material for hats, thereby saving the Beaver from extinction. Abraham Lincoln's stovepipe hat was silk. But the three-hundred-year search for Beaver, to use in the fashion industry, was responsible for the exploration and opening up of much of Canada and the northern United States, including our own Boundary Waters–Quetico land.

Timberrrr!

A busy Beaver can chew through a 2 1/2-inch diameter aspen in thirty seconds. Known to fell huge 28-inch-diameter Cottonwoods in desperate situations, beaver prefer trees 2 to 7 inches in diameter. But they cannot cause the tree to fall the way they want, as evidenced by tree-smashed Beavers. They simply chew around and around the trunk until the tree begins to sway, at which time they run away as fast as their little beaver feet will take them. Trees often do fall toward the water, since leaf growth is heavier on the open and sunny lake side.

Beaver get both food and shelter from felling a tree. Chewing trees replaces a trip to the dentist, too; rodent incisors grow constantly and need to be worn down by gnawing.

Quaking Aspen bark, twigs, and leaves are their favorite food. One acre of aspen will feed a colony of seven Beavers for one year. Alder, willow, birch, Black Ash, Pin Cherry, Mountain Maple, and Beaked Hazel are distant seconds.

Beavers use uneaten tree parts for the construction of dams

and lodges. The sound of running water apparently triggers the dam-building urge. One motivated colony constructed a 2140-foot dam near Three Forks, Montana. The dam creates a suitable pond for the beavers to live in. Recent research has shown that these ponds have an acid-neutralizing effect on the streams that flow through them.

Lodges have two underwater entrances and are normally 5 to 6 feet high and 12 feet across at water level. Stick walls 4 feet thick keep the inner chamber above freezing even in the coldest of months. A good blanket of snow adds to the lodge's insulative value—earth-sheltered efficiency at its best.

Rodent brains

French biologist P. Richard set up several fascinating experiments to test the creative intelligence of several rodent species. In one test, a piece of bread was suspended from a string that was anchored at the ground. The rat and Muskrat lept for the bread in vain, while the Beaver simply chewed through the anchor string to get the bread. Beaver—1. Other rodents—0. In the second test, a succulent willow was protected by wire fencing at the base. The Beaver simply stacked debris until it could climb over the fence and chew on the tree.

Sparky says: Read up on the fur trade and its colorful laborers, the voyageurs. Learn their songs (*Allouette, En roullant,* etc.), make thick pea stew, take a paddling break to smoke your pipe, or reenact the "hivernant ceremony" after crossing the Height of Land Portage between North and South Lakes. Visit spots in the Boundary Waters–Quetico mentioned by fur trade journalists. Excellent books include *Where Two Worlds Meet* by Carolyn Gilman, *Portage into the Past* by J. Arnold Bolz, and *Caesars of the Wilderness* by Peter C. Newman.

Treats

Lily pads rolled neatly into a cigar shape make a true summer delicacy for the Beaver.

Red-backed Vole

Clethrionomys gapperi

Mouse-sized rodent

Length 3 to 4 1/2 inches

Tiny ears and short tail

Reddish back

Red-backed Voles are to boreal predators what M&Ms are to hungry office workers: snack food. Everyone eats them. Garter Snakes, Short-tailed Shrews, Least Weasels, Short-tailed Weasels, Long-tailed Weasels, Mink, Common Ravens, Saw-whet Owls, Boreal Owls, Great Gray Owls, Northern Hawk-Owls, Red Foxes, and Coyotes all feast on the little fellas. The Pine Marten, the king of the vole hunters, subsists almost exclusively on this rodent during peak years of the vole population cycle.

Human visitors often see voles scurrying from under one log to another in the dense Sphagnum Moss mat of the spruce-fir forest. The tiny mouselike rodents you see in the daytime forest are almost certainly Red-backed Voles.

If a Red-backed Vole pauses long enough for you to study it, you'll easily notice the red back contrasting with browner sides. The vole is 3 to 4 1/2 inches long, with a tail that is half as long. It differs externally from mice by having tiny ears and a much shorter tail.

The life and times of the Red-backed Vole

Don't be too surprised to find a Red-backed Vole up a tree without a ladder. They are adept climbers, occasionally even nesting in trees. One vole was discovered nesting 19 feet up! But they have also been known to kill trees up to one foot in diameter by gnaw-feeding on the bark. After a meal like that anybody would need a drink. Red-backed Voles must drink half their weight in water every day. Besides tree bark, they eat seeds, fungi, berries, herbaceous leaves and shoots, and a few insects for good measure. Nonhibernators, they cache away seeds for their wide-awake winter. Occasionally they sneak over to neighbor Red Squirrel's "pantry" and "borrow" some cone seeds for their own use.

A critter that makes such a popular snack for the predators must also be able to produce batch after batch of its own kind. And indeed the Red-backed Vole is prolific, able to produce four to six young every three weeks from late winter through fall. They occasionally practice postpartum mating, which means that within twelve hours after giving birth the female is mating with another male. Young are weaned at seventeen days and gestation is seventeen to nineteen days. Now that's cutting it close. In fact, a case has been reported of a litter being born seventeen days after the last litter, who were still suckling. The mother quickly weaned the last batch and got on with the next! Fortunately, rodent hoards are kept in check by the hungry mouths and beaks just waiting outside the security of the den for a little snack . . . a "Tootsie Vole" if you will.

 Sparky says: Get on all fours and crawl ever so slowly through the deep moss of the boreal forest, your face only inches from the ground. Poke your nose into holes. Smell the earth. Take a magnifying glass along and notice the tiny details: moss fronds and spore caps, scattered seeds, lichens, and insects. You'll experience the world of a Red-backed Vole.

White-tailed Deer

Odocoileus virginianus

Adult deer average 3 feet high at shoulder and weigh 100 to 300 pounds

Coat is gray in winter and reddish in summer

Bucks grow branched antlers in summer and drop them in mid-winter

The logger, deer hunter, and Timber Wolf can be three of the best friends a White-tailed Deer ever had. By removing canopy trees, logging opens the forest floor to sunlight and abundant new growth, which deer relish. This plentiful food source allows the deer to thrive and multiply. If unchecked, a healthy herd can nearly double its numbers in one favorable year. The result is overgrazing, starvation, and death. Starvation is a terrible way to die. Enter the predators—human and wolf. Controlled hunting by human and natural predation by Timber Wolves culls the weak, thins the herd, and creates a vigorous, healthy population that can survive on the available resources.

Before the turn-of-the-century lumberjack arrived, much of the canoe country was old-growth spruce-fir forest, and "deer-less." The wilderness we now call the Boundary Waters and Quetico hardly knew the delicate step of a White-tailed Deer. It was Moose and Woodland Caribou country. But by the 1930s logging and subsequent brush fires created ideal Whitetail habitat. They flourished, nourished by the new growth, then starved as it was eaten up. Today, with the suppression of fires and protection from logging, the BWCAW–Quetico forests are maturing and the White-tailed Deer population is decreasing.

Bucks often exceed 36 inches at the shoulder and can top 250 pounds in weight. The Minnesota record, a Cook County specimen, weighed 400 pounds dressed out, and alive may have exceeded 500 pounds. Does rarely weigh more than 150 pounds. A red summer coat gives way in fall to the thick gray hollow-haired insulating coat of winter. When fleeing, White-tails send up the danger flag, the erect white tail and rump that alerts others to possible trouble.

North Woods smorgasbord

Have you ever noticed the neatly manicured shores of some lakes? The cedars seem to have been precisely pruned of all their greenery below 7 feet. What's responsible? Ice? No. It is the work of wintering Whitetails who have browsed their favorite trees as high as they could reach. Once the snow is belly deep, travel becomes difficult, and deer yard up in dense cedar stands, where they share well-packed trails. Deer yards provide food and protection from winter's icy blasts. Cedar, though a nutritious and preferred winter browse, does not grow enough over the summer to provide an abundant cold-season supply. Whitetails instead subsist on 6 to 8 pounds per day of Red Osier Dogwood, Mountain Maple, and Beaked Hazel twigs. Malnutrition and starvation eventually results from such a nutrient-poor diet if winter lingers.

Spring comes just in time for most winter-stressed, fat-depleted deer. Snow-bare south-facing slopes provide the ever-green leaves of small plants such as Bunchberry, Wintergreen, and Wild Strawberry. As spring progresses, deer gorge on the highly nutritious and easily digestible lush grass of green-up. In late spring and early summer, Whitetails wade belly deep in lake margins, feasting on a variety of aquatic plants, that are high in sodium. (Some water plants have a thousand times more sodium than land plants.) No low-salt diet for these guys.

Young aspen leaves from suckers less than a year old are favored in summer. The abundance of such suckers makes newly logged and burned sites very attractive to deer. They eat leaves of maple, birch, willow, cherry, honeysuckle, rose, and strawberry as well. Fruit and mushrooms provide variety.

The only deciduous green leaf that stays green well into fall is that of Large-leafed Aster. Deer relish it. Come late fall, deer again switch to grasses, sedges, and small evergreen plants. Woody browse (bark and twigs) is the last, but necessary, resort as winter sets in and the cycle repeats.

A diverse diet is crucial to the White-tailed Deer. Deer have the uncanny knack of choosing the most nutritious food available to them. Researchers B. L. Dahlberg and R. C. Guettinger found that deer fed a winter diet of straight cedar fared worse than those subsisting on a variety of second-choice

Antler-eating Mice

"Why don't I ever find deer antlers in the woods?" If I told you that little mice ate them, would you believe me? Voles, mice, and squirrels gnaw on winter-shed antlers to get precious calcium into their little bodies. Gnawing also keeps these rodents' constantly growing teeth in check. Even calcium-starved deer will gnaw on their own shed antlers.

woody browse. A varied menu of seminutritious browse evidently limits their intake of certain plant compounds that would inhibit digestion if eaten in large quantities. This explains why deer fed only cedar lost more weight than those on a lower-nutrient but varied diet.

Sparky says: Find a muddy piece of shoreline and practice your tracking skills. The Whitetail's heart-shaped hoof track always points in the direction of travel. Can you find a similarly shaped but much larger Moose track, or the webbed footprint of a Herring Gull? Bring a bag of plaster of Paris along to make a track-cast souvenir of your trip. Mix the plaster with water until it is the thickness of pancake batter, then pour gently into the track. A cardboard ring around the track will help contain the plaster. Let it dry until very hard, and gently remove with the help of a knife. Brush it off and rinse with water. The raised plaster print will actually be a model of the animal's foot.

The largest mammal in the canoe country may top 1200 pounds, yet thrives on a diet of water plants and twigs. In fact, the word *moose* means "twig eater" in the Algonquin dialect of the Ojibwa language. The twigs of willow, aspen, birch, Mountain Maple, Beaked Hazel, American Yew, Red-Osier Dogwood, and Balsam Fir provide the preferred winter browse for this humongous herbivore.

Most summer canoe country paddlers encounter Moose by interrupting their dinnertime. Knee deep, belly deep, or totally submerged along lake edges, Moose feed on water plants such as pond weed, bur reed, and water lilies. The canoeist can get close by making short paddle bursts every time the Moose dunks his head for another bite of lily pad.

Boundary Waters–Quetico visitors regularly encounter swimming Moose. It is amazing that a half-ton animal can dog-paddle its way across a large lake at a good clip with only four long skinny legs to work with. Researchers believe that in the early 1900s Moose swam the 15 miles from Ontario to colonize Lake Superior's Isle Royale. But never try to hitchhike a ride on the back of a swimming Moose, as several northern guides have claimed to do. If the Moose makes it to shore before you slip off, you'll surely become vulture pickings.

North America's largest member of the deer family, Moose are conspecific (same species) with European Moose, which are known as "elk" there. The average man cannot look over the shoulders of a bull Moose. Yet these massive animals can gallop at 35 miles per hour and finesse their way through thick brush. Cows are three-quarters the size of bulls. Both have large shoulders, long legs, a huge snout, and a pendant, or "bell," hanging below the throat. Some say they look ungainly and homely, and I'd have to agree.

Bulls begin growing antlers in April. The velvet, which supplies blood to the developing bone, is rubbed off by August, leaving large, palmate-tined antlers. These racks may spread 7 feet across and weigh 75 pounds. If it weren't for mighty neck muscles, bull Moose would be dragging their racks through the

Moose
Alces alces

Bulls may top 1200 pounds and stand 6 feet high at the shoulder

Antlers of bulls are large and palmate

Huge snout and gangly legs

Road lickers

Kneeling Moose pose a serious winter driving hazard in the northland. Sodium-starved Moose eagerly lick up salt grime dropped by passing cars. Interestingly, Moose only kneel with their front legs, while the hind legs remain standing.

Stomach capacity

The Moose stomach can hold 112 pounds of food at one time, which comes in handy for a mammal that needs 50 to 60 pounds of browse per day.

moss. The racks become battle tools during the September–October rut. Antlers are usually shed in January and February, although I've seen antlered Moose as late as February 16.

Moose sign is abundant in the canoe country. Huge heart-shaped tracks always point in the Moose's direction of travel. Look also for "Moose marbles," marshmallow-size brown droppings deposited in piles.

Itchy backs and fuzzy brains

We've all had itchy inaccessible spots on our body that feel soooo good when they get scratched. This scratching relief nearly got Minnesota's Moose population in serious trouble recently. Thriving populations of Winter Ticks, or Moose Ticks, overran them in the winter of 1990–91, sucking blood from all parts of the Moose's body. The only relief was a good rub on a rough-barked tree. But along with the ticks, off came big patches of their hollow insulating hair. Hypothermia killed many. Come spring, infant survival was adversely affected by stress-weakened cows. The Arrowhead region of Minnesota may have lost 50 percent of its Moose population over three winters, 1988 to 1991. Wildlife managers were so concerned that they closed the fall 1991 Moose hunt in northeast Minnesota. But biologists are optimistic about the future. This Winter Tick explosion was exceptional, with unique conditions leading up to it. Future tick cycles, though inevitable, are likely to be less severe.

If that isn't enough to worry about, canoe country Moose also have to beware of their smaller cousins, the White-tailed Deer, who can pass on the fatal neurological disease known as Moose Brainworm. Deer pass the eggs of a parasitic flatworm in their droppings. Feces dropped in the water allow the eggs to develop and reproduce in a species of aquatic snail. The immature worm leaves the snail and may be slurped up by a drinking deer or Moose. It then work its way to the surface of the brain. That's where it stays in Whitetails, but in Moose it burrows into the brain, wreaking havoc on the "control center." A brainworm victim becomes confused and often wanders far from home, recklessly encountering humans and urban/suburban jungles.

Fortunately, Moose and Whitetail ranges don't usually overlap. Moose prefer older mature woods, while Whitetails thrive in cutover new growth forests. American Yew and Red Osier Dogwood are the only common food preferences.

 Sparky says: I saw a Canadian Cree Moose hunter demonstrate this trick on television once. Find a swampy shore covered with Moose tracks. Take a birch-bark Moose call (or a camp pot), quietly fill it with water, then pour it slowly and evenly into the shallow water. Wait. You've just imitated the sound of a cow Moose urinating, and if you do this during the autumn rut, you soon may be nostril to nostril with an excited bull. Caution is advised.

Green hair

Moose have green hair between their toes. Pheromones secreted during the rut cause the staining.

Mammals of the Boundary Waters and Quetico

Family Talpidae: Moles
- ❏ Star-nosed Mole — *Condylura cristata*

Family Sorcidae: Shrews
- ❏ Masked Shrew — *Sorex cinereus*
- ❏ Arctic Shrew — *Sorex arcticus*
- ❏ Northern Water Shrew — *Sorex palustris*
- ❏ Pygmy Shrew — *Microsorex hoyi*
- ❏ Short-tailed Shrew — *Blarina brevicauda*

Family Vespertilionidae: Bats
- ❏ Little Brown Bat — *Myotis lucifugus*
- ❏ Keen's Myotis — *Myotis keenii*
- ❏ Silver-haired Bat — *Lasionycteris noctivagans*
- ❏ Big Brown Bat — *Eptesicus fuscus*
- ❏ Red Bat — *Lasiurus borealis*
- ❏ Hoary Bat — *Lasiurus cinereus*

Family Leporidae: Hares and Rabbits
- ❏ Snowshoe Hare — *Lepus americanus*

Family Sciuridae: Squirrels
- ❏ Eastern Chipmunk — *Tamias striatus*
- ❏ Least Chipmunk — *Eutamias minimus*
- ❏ Woodchuck — *Marmota monax*
- ❏ Red Squirrel — *Tamiasciurus hudsonicus*
- ❏ Northern Flying Squirrel — *Glaucomys sabrinus*

Family Castoridae: Beaver
- ❏ Beaver — *Castor canadensis*

Family Cricetidae: New World Rats, Mice, Voles and Lemmings
- ❏ Woodland Deer Mouse — *Peromyscus maniculatus gracilis*
- ❏ Red-backed Vole — *Clethrionomys gapperi*
- ❏ Heather Vole — *Phenacomys intermedius*
- ❏ Meadow Vole — *Microtus pennsylvanicus*
- ❏ Rock Vole — *Microtus chrotorrhinus*
- ❏ Muskrat — *Ondatra zibethica*
- ❏ Southern Bog Lemming — *Synaptomys cooperi*

Family Zapodidae: Jumping Mice
- ❑ Meadow Jumping Mouse   *Zapus hudsonius*
- ❑ Woodland Jumping Mouse   *Napaeozapus insignis*

Family Canidae: Dogs
- ❑ Coyote   *Canis latrans*
- ❑ Timber Wolf   *Canis lupus*
- ❑ Red Fox   *Vulpes vulpes*

Family Ursidae: Bears
- ❑ Black Bear   *Ursus americanus*

Family Mustelidae: Weasels and allies
- ❑ Pine Marten   *Martes americana*
- ❑ Fisher   *Martes pennanti*
- ❑ Short-tailed Weasel, Ermine   *Mustela erminea*
- ❑ Mink   *Mustela vison*
- ❑ Wolverine   *Gulo gulo*
- ❑ River Otter   *Lutra canadensis*

Family Felidae: Cats
- ❑ Mountain Lion, Cougar   *Felis concolor*
- ❑ Lynx   *Lynx canadensis*
- ❑ Bobcat   *Lynx rufus*

Family Cervidae: Deer
- ❑ White-tailed Deer   *Odocoileus virginianus*
- ❑ Moose   *Alces alces*

Birds

Common Loon

Gavia immer

Length 2 feet

Large black and white diving bird

Red eyes

Strange yodeling, wailing, and laughing calls

If it weren't for the Common Loon, most of the North Country's gift shops would be giftless. Loon slippers, loon napkins, loon bookends, loon posters, loon carvings, loon coasters, and loon sweatshirts abound. What is the allure?

The loon is a large, beautifully plumaged bird that inhabits northern wilds and is often seen by canoe country campers. But there is much more to the Common Loon than meets the eye. It captures the imagination of anyone who hears its plaintive cry or wild laugh piercing the darkness over a still northern lake—a mesmerizing experience, even a spiritual one.

Stark black and white plumage, red eyes, and a black daggerlike bill mark the bird. Males and females are indistinguishable in the field, unless you see a male on top during mating or a female laying an egg. The loon's large size surprises many. A loon's swimming body is 2 feet long; in flight it stretches to 3 feet from bill tip to trailing feet. The loon has a 5-foot wingspan.

The name "loon" may have originated from the Old Norwegian *løm*, which means awkward or clumsy. This likely refers to the loon's inability to walk upright on land due to feet placed far back on the body. The English word "lummox" may also have its etymological roots in *løm*. But "loony" does not refer to the insane laugh of a calling loon; it is derived from the word "lunatic."

Some people propose that the common names of birds be standardized around the world. The British call the Common Loon the Great Northern Diver, which is descriptive and accurate but lacks the pizazz of "loon." I prefer our name.

Historically, the Ojibwa called this bird *mahng*. In the Ojibwa creation story, Mahng dives deep into the water and succeeds in raising part of the flooded earth after Otter, Beaver, and Muskrat have failed. Mahng was one of the largest totems, or clans. People of the mahng totem claimed to be of the chief family, and pointed to the loon's "necklace" as proof. The necklace resembled the *megis* (seashell) necklace worn exclusively by chiefs.

Let's learn loon language

Common Loons use four distinct calls to communicate.

Yodel—a sustained call of several three-syllable squeals. Given only by males, this yodel is the loon's true song, which attracts a mate and defends a territory. Sonograms reveal that each male has a distinctive voice, easily identified by neighboring loons.

Wail—a long, mournful cry similar to, and often confused with, a wolf howl. Wolves, Coyotes, and dogs have been known to answer the wail. This call is the basis for inter-loon communication. According to loon researcher Judith McIntyre, it can be interpreted as "come here" and "here I come." It brings mated pairs together and calls the chicks back to the parents.

Tremolo—a quavering laugh. This distress call is given when danger threatens—for example, when humans canoe too close to chicks or a nest. The tremolo is the only call given in flight. (Maybe loons have a fear of flying!)

Hoot—a soft one-note call loons use to talk to their close neighbors. "Hi, how are you? I'm fine, thanks." They also use it to keep tabs on the kids. "Stay close now, little ones." Canoeing loon linguists have to listen closely to hear this one. The hoot is the only call used by loons on their wintering grounds.

Not all loon language is vocal. In one of the most-asked-about behaviors, a loon draws its body upright, neck coiled, wings stiffly spread, and charges a canoe. The tremolo often accompanies this act. As you might guess, this is a highly aggressive act, usually in defense of young. The conscientious canoeist will paddle on.

Divinely designed diving machine

This bird is made for diving, not for flying. Hollow bones allow most birds to take flight with ease. But have you ever seen

a hollow-boned Mallard try to dive? It immediately pops to the surface. Loons have solid bones, which increase their density so they don't have to fight buoyancy to stay underwater. Air sacs under the skin and air trapped in the feathers are the only things that keep loons afloat. They can slowly release this air and submerge quietly like a submarine, but when front diving they empty the air sacs quickly.

Before diving, a loon performs some surface reconnaissance. Swimming about with head submerged, it peers into the depths with red eyes, searching for a meal. Red eyes may help underwater vision. The theory is that 10 to 15 feet underwater, the red color of the spectrum is filtered out and appears gray. If the eye were any other color, both that color and red would be eliminated, reducing available light for fishing.

The average summer feeding dive lasts 42 seconds. Loon observers often report much longer dives, but most likely the loon has quietly surfaced and sunk several times before being spotted. Lake Superior fishermen claim to have found Common Loons caught in nets 240 feet below the surface. Solid bones come in handy at these depths, where the pressure is 106 pounds per square inch—7.5 times greater than at the surface.

Large (4 inch by 5 inch) webbed feet propel the loon at great speeds and to great depths. Feet are the loon's motor. And where is the best place to put a motor? On the back of the boat, of course. That is exactly where the loon's feet are located—far back on the body. The loon doesn't use its wings underwater but holds them close to the body, maintaining a hydrodynamic water-piercing shape.

A loon undergoes a 90 percent metabolic reduction underwater. Its heart rate slows and oxygen is carried only to essential organs and tissues. Large amounts of myoglobin in the loon's muscles allow it to store great amounts of oxygen. The oxygen can be chemically extracted when the loon's body needs it, especially during an extended dive for fish.

Size and species of prey seem to matter little to Common Loons. Researcher Judith McIntyre watched one loon suck in

Equality

Both parents incubate the eggs, and for equal amounts of time. To quote naturalist Denny Olson, "It's like men being pregnant for 4 1/2 months."

124 Fathead Minnows in a single feeding session. Common Loons can also swallow whole fish that weigh well over a pound, though one gluttonous loon choked to death on a 2-pound Walleye. Its red eyes were bigger than its stomach. A pair of Common Loons with two chicks consumes 2310 pounds of fish during the fifteen-week chick-rearing period.

Yellow Perch are the most common food for loons in Ontario, Minnesota, Wisconsin, and Michigan. The perch's zigzagging swim in the upper lake layers makes it easy to see and follow. Loons also take trout, but that fish moves straight and fast and heads for the dark depths when pursued. But loons are opportunistic and will supplement their fish diet with crayfish, aquatic insects, leeches, frogs, and mollusks. On fishless lakes in British Columbia, researcher J. A. Munro observed loons surviving and thriving on a totally fish-free diet.

108 miles per hour

When you're built for diving, flying becomes a major chore. Common Loons weigh as much as Bald Eagles but have only half the wing surface area. In fact, loons have the smallest ratio of wing surface area to body weight of any flying bird. Running into the wind, they may require as much as a quarter mile (or as little as 60 feet) of water runway to get airborne. That's why you'll never see loons on tiny little beaver ponds. On a calm day, after a huge fish meal, a loon may be unable to take off at all. But once it's flying, watch out. With wings beating 260 to 270 times per minute, the heavy-bodied loon can cruise at an average air speed of 75 miles per hour! Sustained flight speeds of 108 miles per hour have been recorded.

Mirages pose a true migrational hazard to loons. Every spring and fall one reads about Common Loons who mistook a rain-covered parking lot for a lake. They land and are stuck, for they cannot get airborne from terra firma.

Salt glands

Boundary Waters and Quetico loons are believed to winter along the south Atlantic coast from North Carolina to the Florida Keys and in the Gulf of Mexico. How does this freshwater bird survive in a saltwater environment? To filter out the salts taken in with 1 quart of seawater, a loon would have to produce 2 quarts of urine. Impossible! Instead, salt glands at the base of the bill remove salts from the blood, and the resulting concentrated solution, which is about 5 percent salt, dribbles out the glands.

The acid rain–mercury connection

In 1983 more than seven thousand wintering Common Loons died in the Gulf of Mexico. The cause of the die-off was a mystery, but the elevated mercury levels found in the loons raised questions. Where did the mercury come from?

Soon after, the Pollution Control Agency found abnormally high levels of mercury in fish from northern Minnesota lakes. Testing of livers from carcasses and feathers from live loons revealed that Minnesota loons possessed mercury levels that had been found to be harmful to Mallards. The source of the mercury and the mechanism of its release into the ecosystem form an ominous cloud over the Common Loon's future.

The theory is that acid rain and acid snow runoff lower the pH level of northern granite-bound lakes. The acidic waters break chemical bonds in the granite releasing naturally occurring mercury into the waters. Mercury passes up the food pyramid from invertebrate to fish to loon. Stored in fat, mercury is harmless. But during the stressful periods of egg-laying and migration, the loon's body metabolizes fat for energy. The mercury then enters the bloodstream and spreads to the vital organs of the body.

A second danger is acid rain's effect on the loon's food base. As a lake's pH level falls, the invertebrates disappear, the fish starve, and the loons are next in line. This tragic story may or may not be reversible. The canoe country without the cry of the loon is quite unimaginable.

Hitchhiking

Loon chicks, or "loonlets," spend 60 percent of their first three weeks of life on a parent's back. Protection from hungry Snapping Turtles, Northern Pike, and cold June waters are good reasons for loonlets to hitch a free ride.

Loons galore

A 1990 Minnesota DNR census found 12,039 Common Loons in the state—more than in all the other lower 47 states combined. Alaska and Canada have healthy loon populations that are too huge to count.

Sparky says: Become a loon interpreter! Refer to the loon language section and learn all four calls. See if you can identify the yodel, tremolo, wail, and hoot on your next trip. Remember, only the male yodels. The hoot is the most difficult one to hear, since it is very soft. Once you master these you'll be an official loon linguist!

The "punk duck"—I've heard it from my campers many times. "Hey, look at that punk duck!" You've seen it too if you've paddled the northern waterways. It's the duck with "bed head" or a bad case of static cling or a punker's haircut. The reddish head feathers of the female stick up and back, giving the Common Merganser a disheveled appearance. Ornithologists call it a crest. Whatever you call it, it gives the merganser a less-than-serious look, even though living and raising a family of merganserettes in the great North Woods is serious business.

Watching a duck fly into a hole in a tree can be an enlightening experience. Many canoeists, who have no idea that some ducks actually nest in unrented woodpecker holes, are astounded. Wood Ducks, Common Goldeneyes, and Common Mergansers all make trees their home of choice in the Boundary Waters–Quetico. A tuft of white down stuck to the edge of the hole may alert you to the presence of a duck nest inside. The female merganser lines the cavity with her own gray down and white breast feathers, lays six to seventeen eggs, and settles down to incubate the clutch. Home, sweet home. The male retreats to backwater rushes and sheds his gaudy plumage. He plays no part in raising the family.

The only thing more shocking than seeing a duck fly into a tree hole may be seeing little fluffball merganserettes fall from the sky. When the time comes for the young to leave the nest hole, the mother encourages, coaxes, nudges, and finally shoves the young out of the hole. Their light weight and downy plumage save them from having a "crushing" experience. Quickly they all waddle down to the safer environment of water. For the next several weeks they play follow-the-leader with their mother, learning all the ins and outs of being a merganser. This is when most canoeists encounter them, the mother out in front and near shore, followed closely by her cluster of ducklings. If you should paddle too close, the entire mass splashes away in a frantic cross-water run.

Common Mergansers are slightly longer than Mallards.

Common Merganser
Mergus merganser

Stream-lined diving duck

Males have green head, bright orange bill, and white body

Female's head feathers are rust colored and shaggy

Bills are long and slender with serrated margins.

The female has a gray body, a white breast, a rust-brown shaggy head, and an orange-red thin pointed bill.

On early summer canoe trips you may encounter the dapper breeding male. He sports a brilliant iridescent green head (no mohawk for him!), a blood red bill, and a gleaming white body. One-year-old males and adult males who have molted their breeding colors resemble the female. This stage is known as eclipse plumage.

During their low rapid flight both male and female reveal a large white patch on the trailing edge of each wing. Males have a longer white patch on the leading edge of the wing. This wing patch is also present in eclipse plumage (molted) males and is the only definitive way of separating males from females at that time of year.

Adult mergansers are known for their low-level cruising. With white-flashing wings nearly touching the water's surface with every beat, they zigzag down winding rivers or zip arrow-straight over lakes, all the while giving low rasping croaks.

The Hooded Merganser also inhabits the canoe country. It is much smaller than the Common Merganser, has an all-dark bill, and prefers beaver ponds and smaller lakes. The adult male has a striking fan-like black-and-white crest.

A fishy story

Mergansers were once shot in the belief that they seriously depleted game fish populations. In fact, they mainly eat minnows, rough fish, crayfish, and frogs. Researchers believe that predators keep fish populations healthy by culling the slow, sick, and old, preventing overpopulation and eventual starvation.

Sawbills

All six species of mergansers are diving birds who depend on fish to fuel their internal fires, and all have tiny serrations edging the bill to aid in catching and holding flip-floppy, slimy fish. A handy feature, indeed, it has given mergansers the nickname "sawbill." They've been called other things, like fish duck, diving goose, goosander, and woozer, but sawbill seems the most appropriate. Early trappers, surveyors, and outfitters liked the name as anyone can attest who has ever approached the BWCAW via the Sawbill Trail, rented a canoe from Sawbill Outfitters, or paddled out to Sawbill Lake.

Sparky says: Check out holes in trees to see who lives there. Common Mergansers nest in holes; carpenter ants, woodpeckers, wolves, bears, turtles, kingfishers, chickadees, sawyer beetles, flying squirrels, voles, and chipmunks also begin life in tree holes, dens, burrows, and tunnels. Armed with a flashlight and long-handled mirror, search out such holes and peer into the dark recesses. Do you see any signs of habitation? Is anybody home? Explore and have fun.

One-room apartment

Common Mergansers prefer old woodpecker holes for nesting but also will lay eggs in abandoned hawk nests, on the ground under overhanging rocks, in dense brush, in riverbanks, or in the recesses of rock piles.

Mallard Duck

Anas platyrhynchos

Black Duck

Anas rubripes

Male mallards in spring/early summer have iridescent green heads

Mallard wing speculums are blue, edged in white

Black Duck speculums are violet, edged in black

It's true. The Mallard begging bread crusts at your wilderness campsite is the same species as the Mallard begging bread crusts at your local urban pond. A truly adaptable duck. The Black Duck, the Mallard's country cousin, prefers its territory a little wilder. Both thrive in the Minnesota/Ontario bush, where they may be seen in nearly any patch of wet. On occasion you may even find one high and dry, far from any water. I remember well my surprise when, while looking for Peregrine Falcon nests on the Grand Portage Indian Reservation, I saw a hen Mallard flush from her nest of eleven eggs located at the top of a rocky slope at the base of a cliff, 300 feet above the nearest pond. I imagine it was a safe place, but those ducklings had a long journey overland to the pond.

The best place to observe Mallards and Black Ducks is from the comfort of your own campsite. It will be obvious at which campsites ducks have been fed. I'll never forget the pure joy this activity gave a group of teenagers from inner-city Chicago, whom I was guiding on the Granite River route. Waterfalls were fascinating, catching frogs was great, but nothing could top the experience of having Black Ducks snatch bits of Rye-Krisp from your own lips! These city kids were giggling with wonder and excitement.

Dabbling ducks feed butts-up. They simply tip up and plunge their head underwater to feed on seeds and aquatic insects. They prefer to dabble in water 16 inches deep or less so they can reach the bottom with their searching bills. Human fingers, with twenty nerve endings per square millimeter are less sensitive than Mallard and Black Duck bill tips, which have twenty-eight nerve endings per square millimeter. Compared to their diving relatives, dabblers have very light bones, smaller feet, and legs located more forward on the body.

The drake Mallard, with his resplendent green head, is well known to most humans. His mate is a mottled brown, with orange legs and an orange bill marked with black. When egg-hatching time comes around, the drake gathers with other males to "skulk in the reeds" as they become flightless due to the molting of their feathers. He emerges looking for all the world like the hen, except for his olive bill. The wings of all Mallards, when sitting and flying, exhibit an iridescent blue speculum bordered in white.

The Black Duck, like a darker version of the hen Mallard, appears all black when in flight. The speculum is violet edged in black. The lack of white on the speculum is the surest way to tell a Black Duck from a Mallard. To differentiate Black Duck drakes and hens, check the bill color. His is yellow, hers dull green.

McBlallard

"McBlallard" is one nickname given to the hybrid offspring of a Mallard and a Black Duck. The male exhibits traits from both species, having the black-edged violet speculum of the Black Duck, a Mallardlike green swath on the head, and a chestnut and black breast. But wait a minute—if two distinct species can produce viable offspring doesn't that mean that they are really one species? Maybe. Herein lies the controversy.

Hybrids seem to be most common on the eastern seaboard but can be found in nearly every population where the species overlap. Mallards and Black Ducks share loafing rights on certain canoe country beaches, so some mixed matings likely result. Mallards have also formed rare hybrids with Northern Pintails and other dabbling species.

Social experience seems to play a major role in determining who mates with whom. To prove this, Lynn Brodsky, C. Davison Ankney, and Darrel Dennis raised duckling Mallards with other duckling Mallards, Black Ducks with Black Ducks, and Mallards with Black Ducks to see if, as adults, they would choose partners of their own species or of the species they grew up with. The results were conclusive as 90 percent chose mates from among the ducks they were raised with.

Mallard speculum

Black Duck speculum

Acid rain, lead poisoning, and blackfly parasites

Emissions of sulfur dioxide (SO_2) and nitrous oxides (NO_x), the principal agents in acid rain formation, have increased by 50 percent since 1955. Over the same time period, Black Ducks have suffered a dramatic 60 percent decline in their population. The link seems to be that the range of the Black Duck overlaps nearly perfectly the range of acid-sensitive lakes in northeastern North America, from the Boundary Waters and Quetico east to Quebec and the Appalachians. In acidic lakes invertebrates die out, so ducks can't ingest protein needed for the critical functions of growth and egg laying. The Black Ducks lay fewer eggs, move on, or die.

Black Ducks, fortunately, seem to be less susceptible than Mallards to the blackfly-borne parasite, *Leucocytozoon anatis*, which causes a malarialike disease in young ducks.

Lead poisoning threatens both species. As dabblers, these ducks tip forward, butts in the air, and feed on seeds and critters from the lake bottom. In the process they also pick up hunter's spent lead shot. Iron may be good for you, but lead is not; death by lead poisoning is slow and agonizing.

Mallard melee

Ornithologists call the phenomenon forced copulation. I witnessed it once in a suburban Minneapolis park. A hen Mallard was pursued by five drakes. In their attempts to mate with the fleeing hen, they tried to run up her back, grab her head feathers with their bill, and mount her.

This activity can hardly be called courtship ritual, because it includes no preliminary displays. Several theories have been put forth to explain the aggressive behavior, including development of tensions in populations with an abnormal density of birds. This may well have been the case in the suburban park. Forced copulation is rare in the nonurban Black Duck.

Frozen eggs

Eggs of Mallards have been known to freeze and crack but still hatch out healthy ducklings.

Quack!

The male Mallard does not "quack." He makes a high, short whistle during courtship and also produces a higher pitched and nasal "rhaaaaeb." The hen does all the "quack"-ing.

Sparky says: Practice your wildlife photography skills on approachable animals. Campsite Mallards and Black Ducks make excellent subjects for the budding wildlife photographer. Any camera will do, since your subjects have little fear of people and can be approached closely. Instead of a simple portrait, why not try to capture on film some aspect of duck behavior, such as butts-up feeding, mother/duckling interaction, or aggression. Experiment with an underwater Kodak disposable camera; share the results with friends. Don't waste film on soaring eagles a mile up or loons half a lake away.

Farm ducks

The domestic fat white farmyard duck was bred from wild Mallards beginning approximately four thousand years ago. Like domestic turkeys, chickens, pigs, and beef cattle, ducks have been bred to be large and fat, with the greatest possible amount of marketable meat on their bones.

Great Blue Heron

Ardea herodias

Three to 4-foot tall wading bird with long legs and a long neck

Body bluish while face is marked with white and black

Flys slowly with neck tucked in and legs trailing behind

Hoping to come upon a Moose, belly deep in water lilies, you paddle around the bend as silently as a mere mortal can . . . in an aluminum canoe. Nine times out of ten it will be a Great Blue Heron, not a Moose, who will greet you. "Greet" is probably not the right word for this encounter, because the large heron immediately flaps away on huge steel blue wings, legs dangling behind. But float on silently and watch, for the Great Blue Heron typically lands only a short distance away and resumes its feeding activities.

Observe it nabbing Northern Pike fry, stalking snakes, or catching crayfish. Ankle deep in water, the heron stands motionless, its body a coiled spring ready to strike. With its head cocked at a tense 45° angle to the water, its eyes intently focus just below the water's surface. A darting sucker or scurrying crayfish triggers the spring; the head plunges into the water, nabbing the unfortunate critter, then flips it into the air and swallows it headfirst. The element of surprise is everything.

The adult Great Blue Heron, in the right light, looks beautifully blue, but at other times it appears only gray. Great Blues make their finest appearance during spring courtship, when long ornate plumes grow on the neck, back, and head. The neck is lighter than the body, and the face is white with a broad black stripe above the eye. The adult's bill is yellow. Male and female have no plumage differences.

Standing next to you, the Great Blue Heron would probably measure somewhere between your belly button and your armpit, but its wingspan is most assuredly larger than yours, at 7 feet. Despite all that surface area, the bird barely tips the scales at 7 pounds. Its light weight and large wings allow the heron to leap into flight and leisurely flap away. But a heron never seems to retract its landing gear, as its long legs dangle behind. The scrunched-up neck and trailing legs are easy field marks to identify a flying Great Blue, even at a distance.

Colonial times

Lucky is the canoe camper who comes upon one of the dozen or so Great Blue Heron colonies scattered about the border country. Bulky stick nests perched precariously high in dead trees give homes to an average of fifty pairs of herons and their offspring. Nests are safely spaced at least two bill lengths from its nearest neighbor. These colonies, which are often located on "predator-resistant" islands, are used year after year during the summer months.

Like vacationers returning to their lake cabin in spring, herons do their housecleaning, by adding a fresh layer of sticks to the nest. Then their minds turn to other things . . . like sex. A couple of gangly 4-foot-tall birds, with wings the size of beach towels, mating 75 feet up on the outermost branch of a dead Tamarack gives new meaning to the word "awkward." But it works, and a few weeks later three to five eggs are laid. Both birds are attentive parents, turning the eggs over every two hours to ensure even development. In a month, out hatch some of the homeliest baby critters ever to live on this planet. (That's from a human perspective you realize.)

Regurgitated fish fresh from a parent's bill makes up breakfast, lunch, and dinner for the fast-growing herons. When a heron returns to the colony from a fishing foray, it utters a hoarse guttural squawk, which its own offspring recognize immediately as the dinner bell. The instant the heron lands at the nest the young grab for its bill. This triggers the parent to upchuck the piscitic puree into their gaping mouths. When the fishing is good all chicks may survive; but if food is hard to come by, the largest, most aggressive young may commandeer all the food and force the weakest sibling from the nest. A Great Blue Heron chick that falls from the nest is doomed, because parents refuse to feed a grounded chick. Turkey Vultures and Red Foxes maintain a death watch from below. At two months the young can fly and fish on their own, and nobody is more relieved than the parents.

Tourist bureau

The colony may act as the local tourist and information bureau. Studies have shown that when one heron finds a hot fishing spot, it may share that information with others in the colony.

The easy way

A Minnesota fisherman laid his nice stringer of Northerns in the trunk of his car. He went back to the dock to get his pole and tackle box, and returned to find a Great Blue Heron picking his fish into bite-size pieces. The heron would not let the fisherman near the car. Hey, do you blame him? It's not often the fishing is that good, or that easy!

Canoeists should steer clear of active heron rookeries and view them with binoculars from a safe distance. Adult herons are easily spooked from the nest, allowing opportunistic Herring Gulls and Common Ravens the time to raid eggs or hatchlings.

Dagger

Beware of the Great Blue's bill. If you should ever happen upon an injured heron, steer clear of its yellow dagger. A case is recorded of one striking at a pine paddle and embedding its bill 2 inches deep.

Sparky says: Put on a "blue" scavenger hunt to find out how common that color is in nature. Starting at your campsite, send everyone off to find natural blue objects. Meet back in a half hour and have each person share a favorite object. If it's a flower or mushroom, leave it intact and bring the whole group to experience it in its natural setting.

Spotted Sandpiper
Actitis macularia

Watch for the Spotted Sandpiper as you paddle along the shore of nearly any lake or stream. Its habit of dashing along the water's edge in short bursts, pausing to grab a protein-rich insect now and then, combined with its compulsive hind-end bobbing, makes the Spotted Sandpiper quite conspicuous.

If you happen to surprise the little sandpiper, it lives up to its name by emitting sharp piping calls as it flutters a short distance down the shore on stiff, vibrating wings. The bird holds its wings straight out and shivers them in very shallow wingbeats with short pauses of gliding. This sequence of flushing, flying, landing, walking, and bobbing may be repeated over and over if you are paddling right along shore.

The Spotted Sandpiper lives up to all aspects of its name, thus flying in the face of the great ornithological tradition of naming birds for their most insignificant trait (just ask the Sharp-shinned Hawk and the Three-toed Woodpecker). The Spotted Sandpiper does have spots on its breast, it does enjoy perusing sandy shores, and it will "pipe" for all it's worth when frightened. Seven and a half inches long, the bird has long legs and a bill designed for probing and picking up insects. The Spotted Sandpiper's color of sand and rock, combined with its random pattern of breast spots, makes it nearly invisible when not moving.

Length 7 1/2 inches

Brown back and bold dark spots on white breast and belly

Feeds along shorelines

Bobs hind end regularly

Bob bob bobbing along

Many birds bob their tails—Palm Warblers, Hermit Thrushes, and White Wagtails do; but very few bob their entire hind ends as the Spotted Sandpiper does. Whenever the legs aren't moving, the hind end is bobbing. This behavior has given rise to the Spotted Sandpiper's nickname, "teeter tail."

What purpose could this behavior serve? H. M. Hall, in his book *A Gathering of Shorebirds,* put forth an interesting theory. He contended that the bobbing motion allows the hunting bird to "blend into the lapping wavelets and the play of light and shadow they create on shore" and therefore to escape the keen eye of a watchful predator. Survival is everything.

Whatever the purpose, bobbing must be a hereditary trait; one researcher recorded a thirty-one-minute-old chick performing the bob. The bird must have a "bob-gene."

Women rule

In 90 percent of the world's nine thousand bird species, the male is the brightly colored one. The male sings and performs extravagant displays to attract a mate, and the male defends a territory only to allow the female her business of laying and brooding eggs and raising the young. But Spotted Sandpipers don't read textbooks. In their world the female displays for the male, initiates breeding, and defends the territory, allowing the male to stay home, sit on the eggs, and, when they hatch, watch the little ones.

Lew Oring and Steven Maxson of the University of North Dakota have done extensive research on a colony of Spotted Sandpipers on Little Pelican Island in Minnesota's Leech Lake. They discovered that Spotted Sandpipers not only exhibit reversed sexual roles but are also polyandrous. In this rare mating system, found in only 1 percent of bird species worldwide, the female mates with more than one male. Now that's real avian liberation!

The females arrive first on the breeding grounds. As males appear, the females begin vigorous courtship displays in hopes of wooing a male to mate with. These courting rituals can get so intense that females may actually engage in wing-to-wing combat. Once mating has taken place, the female lays four eggs in a slight depression scraped in the ground. But motherhood has no appeal for this bird; she steps aside to let the male take over. He quickly settles on the eggs and begins the required three-week incubation. Hatching brings forth four precocial fuzzballs who make the male a very busy single parent.

In the meantime the female has most likely gone on to

mate with and lay eggs for another male . . . or two . . . or three. One observer determined that five females had mated and laid eggs for twelve nests. Of these, two females were monogamous, two had two mates, and one mated with four males during the six-to-seven-week breeding season. But competition for males becomes ever more fierce as more and more males are taken out of circulation by the duties of fatherhood. By the second half of the breeding season on Leech Lake's Little Pelican Island, six or seven females competed for every available male. As expected, the "experienced" females were more successful in the mating game. But a woman's work is never done—the female Spotted Sandpiper must defend a large territory, encompassing the individual territories of all her mated males.

All of this activity is perfectly timed to the abundance of the Spotted Sandpiper's staple food, adult flying insects, either aquatic or terrestrial. Breeding ends abruptly in early July so that young hatched near the end of July will still be able to find abundant insects and grow to full maturity before flying south to wintering grounds in Cuba, in the Amazon Basin of Brazil, or even along a mountain stream high in the Andes.

Sparky says: Test a theory about hind-end bobbing. One researcher believes that the Spotted Sandpiper's hind-end bobbing allows it to blend more harmoniously with its feeding environment amongst the wavelets of the lake shore. Test this hypothesis for yourself by recording bobbing rates in calm and windy weather.

Herring Gull

Larus argentatus

Large gull with gray back and black wing tips

Legs pink

Yellow bill marked with red spot on lower mandible

Identity crisis

If you've ever been confused by gull identification, take heart—the gulls seem to be too. Herring Gulls, for example, have cross-mated with Lesser Black-backed Gulls, Great Black-backed Gulls, Glaucous Gulls, Glaucous-winged Gulls, and Slaty-backed Gulls.

Right off the bat, I want you to know my pet peeve. In the entire universe as we know it, there is no such thing as a "seagull." We have Ring-billed Gulls and Slaty-backed Gulls, Ivory Gulls, and Iceland Gulls, and even a gull named for Napoleon's younger brother, ornithologist Charles Lucian Jules Laurent Bonaparte, the Bonaparte's Gull, but not a single one called a "Sea Gull." Let's not add to the world's confusion by calling all the gulls in the canoe country "seagulls." They are Herring Gulls.

The Herring Gull is the only gull nesting in the Boundary Waters and Quetico. On nearby Lake Superior they nest in large colonies on rock islands, but throughout the canoe country they prefer the solitary isolation of their own rock islet. One island equals one nest. The Herring Gull likes its privacy, as you will quickly realize when you are dive-bombed by a protective parent after paddling too close to its nest rock.

Since the Herring Gull is the only gull species to inhabit the canoe country during summer, identification is simple. Any "seagull" is a Herring Gull. An adult has pink legs, a yellow bill with a red spot on the lower mandible, a white body with a gray back, and gray wings tipped in black. But it's a long journey to reach adulthood and acquire this neat, clean plumage; Herring Gulls must survive three full years of youth in sooty brown feathers before coming of age in their fourth spring. The Herring Gull, therefore, is known as a four-year gull.

Light bulbs in the nest

Ethology, the study of animal behavior in natural settings, was first popularized by Nobel Prize–winning ethologists Konrad Lorenz and Niko Tinbergen. Tinbergen studied the behavior of Herring Gulls for many years, culminating in the publishing of his classic *The Herring Gull's World.* In one experiment he discovered that Herring Gulls are much more attached to the nest than to their own eggs. One gull even sat on an empty nest while its own eggs lay on the ground a short distance away. A parent gull indiscriminately incubated wood eggs, blue-and-yellow painted eggs, light bulbs, and other Herring Gulls' eggs placed in its nest. It drew the line at square wooden eggs, though. Tinbergen postulates that gulls have never developed a mechanism for identifying their own eggs because eggs are ordinarily stationary, staying right there in the nest where they were laid.

The North Woods' brief summer offers some very hot days, and one of the hottest places around is exactly where these gulls nest, on exposed granite rock. To protect themselves from the heat, incubating gulls rotate to face the sun throughout the day, minimizing the total body area exposed to the sun and presenting only their most reflective, and hence coolest, colors, the white head and breast, to the glaring sun.

The chicks hatch with eyes open and a full coat of down. They can walk, but stick close to home for two to three weeks. The parents, who are monogamous, faithfully tend their growing chicks, delivering countless gullet-loads of partially digested fish to them. But the chicks must know the "password" in order to get the delivery. "Open sesame" won't work, but a tug at the red spot on the adult's bill does it every time, triggering the adult to vomit up the chick's meal. Adults continue to feed their young in this manner until they can fly.

Better than slippers

Have you ever seen gulls standing on ice-covered lakes or swimming about in frigid waters and wondered about how they keep their naked little feet and legs from freezing? I have. The answer seems to be in their ingenious circulatory system. In the legs, veins and arteries lie in contact with each other. As warm arterial blood leaves the heart on its journey to the feet, it passes the colder returning venous blood and warms it. Gulls can also constrict the blood vessels to reduce the flow, which reduces heat loss. The body temperature of a gull in 35° F water

Old gulls

Herring Gulls appear to mate for life, with pairs returning to the same nesting site year after year. Some pairs could, theoretically, celebrate a 40th anniversary, as a pair of captive gulls in Morehead City, North Carolina, lived to 45 and 49 years respectively. The "wild" age record is of a 31-year-old gull. Most don't make it to such a ripe old age.

What fish?

Beware of leaving that stringer of fish you caught for breakfast in the water overnight. If a Snapping Turtle or Mink doesn't get them, the Herring Gulls will.

may be 104° F while its leg temperature may only be slightly above that of the water.

 Sparky says: Locate and observe a Herring Gull pair's nest rock. Does the adult sitting on the nest rotate its body to face into the sun on hot days? At a nest with newly hatched chicks, do the adults serve as living umbrellas for the young on hot days or rainy days? Paddle nearby to get a firsthand look at how these gulls protect their nest. Are they more vigorous in defense of eggs or chicks?

I t's a stellar day in the canoe country. The air is calm and warm, the sky is blue and little flat-bottomed cumulus clouds have started to form. Then the bowsman spots it. "Look, up in the sky . . . it's a bird, it's a plane, it's . . . wait . . . no . . . it *is* a bird." A large black bird soaring on upturned wings glides over the canoeists, twitching not one barbule of one barb of one vane of one rachis of one feather. It appears headless but reveals a long tail as it soars out of sight with wings still held in a V.

Meet the Turkey Vulture. It is known to taxonomists as *Cathartes aura,* the "scavenger of the breeze," which is an appropriate name for a bird who rides thermals over large areas, looking for dead critters to devour. And that is most likely how you will encounter the "TV." But don't look for the Turkey Vulture to come gliding out of the morning mists; it needs the thermals created by sun-warmed earth to provide lift for its effortless flight. Afternoon is "vulture time."

The Turkey Vulture will win no avian beauty awards, but it has a perfect design for life as a member of nature's cleanup crew. The red-skinned head has no feathers. This means that the vulture can stick its head deep inside an animal's carcass without caking its head feathers with blood and guts. The bird can't reach the feathers on its head, and so can't clean them. Bald is beautiful.

Turkey Vultures hold their wings in a dihedral when soaring—the shape of a shallow V. You can notice this especially when you view the bird head on or from behind. Bald Eagles and Common Ravens hold their wings straight out in a horizontal plane when riding the thermals. Both of these species also show a neck and head when soaring, while the Turkey Vulture appears headless. All body feathers are black; but when the bird glides overhead, displaying its massive 6-foot wingspan, its flight feathers appear silver gray, contrasting with black wing linings.

Turkey Vulture
Cathartes aura

Huge black bird

Wingspan 6 feet

Head red and featherless

Appears headless in flight with long tail

Soars with wings held in a V

Osteoporosis

Most birds' mineral requirements resemble our own. Calcium is especially important for egg production, and breeding female birds need it in large quantities. But since the bulk of their diet is calcium-poor dead animal parts, vultures must supplement their diet with calcium. They do this by swallowing mice and shrews whole, ingesting the calcium-rich bones.

Avian vitamin needs are also much like ours. But how do carnivores and scavengers like the Turkey Vulture get their "U.S. Recommended Daily Allowance" of vitamin C? They can't gulp a glass of orange juice, but they can manufacture vitamin C in their liver or kidneys.

M-m-m-m, smell that food!

Birds in general have poor olfactory senses. Vision is excellent, hearing is superb, but the sense of smell is often poorly developed. Unlike mammals, who use smell extensively, birds detect food and predators mainly through visual and auditory clues.

Some of the earliest experiments on Turkey Vultures' sense of smell were conducted in the 1920s by Frank Chapman on Barro Colorado Island, Panama. He covered mammal carcasses to make them completely invisible to soaring Turkey Vultures. With no visual clues, the vultures nonetheless arrived on the scene after the decay process had rendered the hidden carcasses quite odoriferous. Rotten, stinking, buried fish elicited no response from the mammal-eating vultures, which further strengthened the belief that Turkey Vultures have a keen and discriminating sense of smell.

These results were duplicated by Oscar Owre and Page Worthington, who used hidden pans of putrefied dog food, horse meat, and freshly killed chicks. These delightful meals were investigated before hidden empty pans 75 to 93 percent of the time. Another study simulated the smell of rotting meat by pumping ethylmercaptan fumes into the air, which did indeed attract wild flying Turkey Vultures. This experiment proved to be valuable to engineers trying to detect leaks in a 42-mile-long pipeline. They simply pumped the same chemical through the line and noted where the Turkey Vultures gathered. One circling flock of vultures equaled one gas leak.

Not everyone is so convinced that vultures have such swell sniffers. It's the old sight-versus-smell controversy. Other experimenters have shown that the threshold for detecting three components of decay is too high to be useful to the high-soaring vultures. But it is quite possible that vultures may be more sensitive to other components of decomposition.

 Sparky says: Watch a Turkey Vulture soar. Watch it closely to see if it flaps its wings . . . ever. Then close your eyes, lie back, and imagine yourself soaring at 300 feet. Imagine the panorama of lakes and trees and sky and tiny, insignificant little human animals scurrying below. Smell the intensely fresh air as you rush through it, hoping to catch a sniff of something rotten. It's a whole new perspective, isn't it?

Bald Eagle

Haliaeetus leucocephalus

Black body with white head and tail

Immature birds are mottled black, brown, and white

Wingspan to 8 feet

Huge bulky stick nest in large trees

Call is squeaky cackle

The Bald Eagle is back! Back from the brink of a DDT-induced extinction to soar over our campsites again. Once considered endangered across the United States and threatened in Minnesota, the Bald Eagle population continues to grow at a rapid pace. Today they occupy niches in the Boundary Waters and Quetico that have gone unused for decades.

Only the California Condor, which may weigh 20 pounds and soar on wings 9 feet across, can top a large female Bald Eagle, weighing 14 pounds with a wingspan of 8 feet, as North America's largest bird of prey. Males, as in most raptor species, are considerably smaller.

You're not likely to mistake the gleaming white head and tail of an adult Bald Eagle as it perches in a shoreline White Pine. But what many people don't realize is that it takes four to five years for these birds to get adult plumage. Until then they are mottled brown with flecks of white under the wings and tail. This leads to confusion with Golden Eagles, which are extremely rare in the canoe country. Immature Bald Eagles show much white on the wing linings, whereas Golden Eagles have dark wing linings.

Don't expect the Bald Eagle soaring over your canoe to let out a majestic scream as they do on the television commercials. That's not their voice! Sound engineers dub the scream of a Red-tailed Hawk over the eagle's own high-pitched squeaky cackling, which apparently is not majestic enough for the producers' liking. (How they got those television eagles to lip synch, I'll never know.)

While soaring at 500 feet, the Bald Eagle can spot a swimming fish a mile distant, swoop down at 60 to 100 miles per hour, grab with its talons, and fly off with 5 pounds of fish. How do they do it? Each eye has two foveae, or retina depressions, (we only have one) that can focus together to give unequaled binocular vision. Bald Eagles' sight is eight times sharper than ours.

The 2-inch-long talons make formidable weapons. If an injured eagle is confronted, it rocks back on its tail and presents the flesh-tearing talons as defense. The beak is very weak by comparison. Under ideal conditions an eagle may be able to lift an 8-to-10-pound fish. But stories such as the September 23, 1929, report of an eagle seizing a 50-pound eight-year-old boy in Somerset, Kentucky, and lifting him 20 feet off the ground are just that—stories. Stories, myths, and lies.

Bald Eagles eat fish and injured ducks, scavenge carcasses, and steal from Ospreys. This last habit, which ornithologists call kleptoparasitism, is a very effective way to secure fish. Noble American symbol or opportunistic scoundrel? That debate has gone on for a long time.

A turkey of an idea

Which would you rather see as the symbol of the United States of America: a turkey or the Bald Eagle?

Immature eagle

On July 4, 1776, a committee of Ben Franklin, John Adams, and Thomas Jefferson was formed to produce a new national seal. The debate raged over what animal would best represent the country. The Continental Congress settled the matter on June 20, 1782, when they chose the Bald Eagle as our national symbol. But not everyone was pleased, least of all Ben Franklin. He called the Bald Eagle a "bird of bad moral character" who didn't "get his living honestly." He was referring, of course, to the eagle's habit of stealing fish from Ospreys. Ben advocated the Wild Turkey, which he praised as being "a much more respectable bird, and withal a true original native of America." I think the Continental Congress made a wise choice. Sorry, Ben.

Dichlorodiphenyltrichloroethane

John James Audubon in 1840 wrote that "eagles . . . have always stirred the imagination of man. They have not, however, always profited by this human interest." Shooting and habitat destruction have taken a toll on Bald Eagle populations, but DDT nearly tolled the death knell for a magnificent species.

In the 1950s, monoculture farming and the desire for increased yields led to the widespread use of chemical pesticides and herbicides such as DDT, a persistent pesticide that concentrates as it passes up through the food pyramid from plankton to minnows to perch to pike to eagle. For example, the plankton may contain only 1/10 of a part per million (ppm) of DDT, while a Bald Eagle on the top of the food pyramid may have a concentration of 20 ppm. That's enough to alter the eagle's calcium metabolism, resulting in abnormally thin eggshells. Eagles, Ospreys, and Peregrine Falcons found scrambled eggs in their nests instead of healthy developing eggs. DDT was banned in 1972 in the United States but continues to be used in other countries.

Since the ban, Bald Eagles have made a remarkable comeback. The North Woods of Minnesota in 1973 contained 115 nests, which fledged 113 young. By 1990, researchers found 435 nests, which produced about one young per nest. That nearly 400 percent population increase in less than twenty years signals growth that is likely to continue.

Year of the eagle

A pair of Bald Eagles climbed high into the fresh air of an April afternoon. He dived and lightly touched her back. They grasped talons and performed a 1000-foot free-fall courtship tumble. Then they broke the grip and met at the huge stick nest wedged into a crotch of an old lightning-struck White Pine. The nest, 6 feet across and 10 feet deep, had been used for twenty-one years, each eagle pair adding a fresh layer of sticks. This pair had been together for seven years. He occasionally brought fresh sprigs of greenery to the nest while she sat on two large bluish white eggs.

A female eaglet hatched in May, and two days later, a male. The female grew faster and larger and gobbled up the lion's

Fresh green pesticides

Ornithologists believe that the fresh sprigs of greenery eagles bring to the nest may have pesticidal properties that discourage parasites such as ticks, mites, fleas, and maggots.

share of the fish, snakes, ducks, and Muskrats that were brought to the nest. The male eaglet starved to death in two weeks.

Chocolate-brown and full grown, she first tried her wings in late June, though the big nest in the old White Pine was still home. By mid-October ancient urges tugged her south. A stiff northwest wind pushed her to the shores of an apparently endless expanse of water. Better to stay over land, she thought. Tiny humans looked up at her from a ridge on the edge of a big city at the end of the lake. A small, agitated black bird harassed her. She soared on.

Winter was easy along the Mississippi River bluffs. With fifty or sixty other eagles she spent the days feeding on stunned fish that had passed through the lock and dam that crossed the wide river. Snow lay deep in the woods and the days were cold, but she had no reason to move south. She had all she needed.

Sparky says: Compare your wingspan to a variety of birds'. Spread your arms wide and have a friend measure the span from fingertip to fingertip. I am willing to wager that nobody can top the Bald Eagle's 8-foot spread. How does your wingspan compare to these other birds'?

Turkey Vulture—6 feet
Osprey—5 feet
Red-tailed Hawk—4 feet
Broad-winged Hawk—3 feet
Sharp-shinned Hawk—2 feet

Two-ton nest

A Vermillion, Ohio, nest occupied for thirty-five years grew to be 12 feet deep and 8 1/2 feet wide. The nest, 85 feet up in a Shagbark Hickory, crashed to the ground during a storm. The massive stick nest weighed 4000 pounds.

Osprey
Pandion haliaetus

Length 2 feet

Black back with white face and undersides

Large fish-eating raptor

Flies with long narrow crooked wings in shallow ΛΛ silhouette

As you paddle down the long sun-sparkled lake, your mind drifts, thinking about other times, other places, and other people. But a movement catches your eye and you snap out of your trance and back into the reality of the present. In a side bay an Osprey hovers above the water as if dangled by an invisible thread, its gangly wings flapping to maintain its position, its eyes intently focused downward, penetrating the water's surface, looking for lunch. Then the thread seems to snap and the Osprey drops straight down with legs extended and plunges feet and face first into the lake. Alas, it was not meant to be on this try, and the Osprey rises laboriously from the water empty taloned, the fish having outmaneuvered the fish hawk.

Not everyone knows *Pandion haliaetus* as the Osprey, for in the not-too-distant past field guides and fishing guides alike labeled it the "fish hawk," a pretty good name for a fish-eating raptor. The French Canadians had the same idea when they called it *l'aigle pêcheur,* or the fishing eagle. But the name I appreciate most is used by the Koyukon people of central Alaska, who call it by the wonderfully descriptive name *taagidzee' aana,* or "one who stares into the water."

No matter by what name you call it, the Osprey is a large hawk, 2 feet long, with a wingspan as wide as an adult human's outstretched arms. The Osprey rarely holds its wings straight out flat, though, but rather glides on crooked wings, which from the front gives the Osprey a flattened ΛΛ silhouette. Viewed from below, the bird is unmistakable, exhibiting large black wrist patches on a background of white. Perched on a limb, the Osprey shows a black back. The black extends onto the white neck and head as a thin band and forms a "robber's mask" across the eyes. The bill is black and the throat, belly, breast, and legs are white.

Double-jointed toes and spiny soles

The Osprey is a piscivore—an animal that makes fish its exclusive food choice. It patrols a regular shoreline beat, flying up and down the lake, peering into the depths, hoping to catch a glimpse of a perch, Smallmouth, whitefish, or anything else with fins. If it spots something, it quickly pulls out of flight mode and shifts into neutral, hovering in place 30 to 100 feet above the water while trying to pinpoint the prey and determine the angle of attack. When all seems right, the Osprey drops out of its hover and plunges feet-first into the lake, its eyes solidly fixed on the swimming target. In this manner it can take fish up to 3 feet below the surface without actually submerging its head and wings. Compact plumage helps reduce the impact and decreases the chances of wetting its feathers and impairing flight.

Sometime between hovering and grasping, the Osprey performs a trick, well known to owls but rare in hawks, as it rotates the outer forward-pointing toe backward. It now has two toes forward and two back, a much better configuration for grasping. Sharp spicules, or spines, on the bottom of its feet also facilitate the death grip. But an Osprey must know its limits and not get too greedy, for a 4-pounder is about all Ospreys can manage to land. Cases have been reported of Ospreys latching onto lunker fish, not being able to release their grip, and being dragged under and drowned. Success is never guaranteed in nature. One researcher, though, reported an 80 percent fishing success rate for Ospreys. I wish I had that kind of luck fishing!

Once caught, the fish is carried head first (always!) in the Osprey's talons back to a perch to eat. Bald Eagles, who are not as skilled in the sport of angling, occasionally may harass flying Ospreys into dropping a catch, then snatch it up for their own. This behavior is known as "kleptoparasitism."

Ospreys have been seen flying low over lakes while dragging their talons in the water. Are they washing off fish slime, trying to attract fish, or simply cooling off? Researchers don't really know.

Long-term lease

One New England nest was occupied every summer for forty-four consecutive years by a pair of Ospreys.

Saved by the oil

Osprey feathers were spared from becoming hat adornment in the millinery trade of the early 1900s because of the oily smell that permeates them.

Breakfast in bed

The female is fed exclusively by the male from pair bonding through egg laying.

Sparky says: Match your skill with the Osprey's. Have you ever tried to catch fish with your feet? Ospreys perform this incredible feat (no pun intended) every day. But we'll give you an advantage and let you use your hands. Locate a school of minnows at the water's edge. Poise quietly above them, then at the right moment plunge your hand lightning-fast into the pool. Did you catch one? No? You now have a new appreciation for the skills of the Osprey.

The Broad-winged Hawk is the most common breeding hawk in the canoe country. You may encounter it while paddling, portaging, or eating a shore lunch. Mated pairs soar in great circles over the thick woods of their territory, giving their high-pitched "p'deeeee" call. Broadwings also give this call while perched deep in the understory of the forest, where they spend much of the summer hunting snakes and frogs. Sitting on a limb partway up a tree overlooking a stream, the Broadwing peers down intently hoping to catch sight of a Red-backed Vole, American Toad, Wood Frog, Garter Snake, crayfish, or large insect. When it does, its tail twitches excitedly and its body sways from side to side before it drops from the perch, talons extended, to grab its prey. One source claims that Broadwings carefully skin their quarry before eating.

Buteos are a subfamily of the hawks that have broad, rounded wings and a short tail, traits that allow them to excel at gliding and soaring. Broad-winged Hawks are small buteos that may reach a length of 16 inches and have a wingspan of 3 feet. "Handsome" properly describes an adult soaring Broadwing, with its white underwings framed in black and its distinctly banded black and white tail. Perched, a Broadwing shows its breast and belly's fine reddish barring. Immatures lack the banded tail, and brown spotting replaces the breast's barring.

As I mentioned above, its song is a very unhawklike high pitched whistle, "p'deeeee." But beware, for the Blue Jay has mastered a Broadwing impersonation and will perform it anywhere and at anytime.

Solar-powered migration

Broadwings make spectacular migrations to and from the jungles of Mexico, Central America, and South America every year. They are one of only thirty-five North American species that winter as far south as central Chile. So maybe they really aren't "our" hawks but rather jungle hawks, spending nine months at home in the tropics and a mere three summering months on

Broad-winged Hawk
Buteo platypterus

Hawk with broad wings spanning 3 feet

Tail is black and white banded

Undersides of wings are white, framed with black margin

Adults show fine reddish barring on chest

Call is a high pitched "p'deeeee."

vacation in the North Woods of Canada and the United States.

Broadwings begin their mass exodus from the northern forests in late August after the young are raised. They make good use of their broad, soaring-type wings by hitching a free ride south on rising warm-air currents called thermals. They will even forsake migrating entirely on cold, cloudy, rainy, "thermal-less" days.

Thermals form as the sun heats up different land forms to varying degrees. Rocks have a higher thermal capacity than trees and therefore heat up faster, radiate the stored heat, and raise the surrounding air temperature. The hot air surrounding the rock floats upward, forming a warm-air bubble that is continuously supplied by rising warm air.

Anyone who has ever canoed the Boundary Waters or Quetico in summer has seen thermals form. It begins late in the morning on clear, calm days, when little, puffy, flat-bottomed clouds begin to form on the horizon. The cumulus clouds continue to form and build throughout the day. The clouds come from warm-air bubbles rising and cooling until the dew point is reached, and the water vapor in the thermals condenses and becomes visible. Voila! Instant clouds!

Broadwings ride these warm-air bubbles on set wings, spiraling higher and higher, using very little energy, until the lift is gone. Then they "stream" off, gliding away from the thermal and losing altitude until they find another thermal to rise on. Thermals can become very large by midafternoon, with one supporting one thousand or more circling hawks. These living tornados, or "kettles," can reach such altitudes as to be invisible to the naked eye. The Broadwings' 3000-plus-mile migration is a long series of thermal-hopping days during which they may flap their wings only in the early morning and on landing for the night. This makes for a fuel-efficient trip, with less time needed for hunting and more time available for "making miles."

Congregations in migration can reach mammoth proportions, as evidenced by the 31,891 broadwings seen over Hawk

Ridge in mid-September 1978. More than 100,000 have been seen in a single day migrating through the narrow Canal Zone area of Panama. Season totals there have topped the 1,000,000 mark!

 **Sparky says:** Combine a fall canoe trip with a visit to Duluth's Hawk Ridge Nature Reserve. The best time to see the peak Broadwing flights is the middle two weeks of September. Ideal migration weather is sunny, with northwest winds that push the inland migrators toward Lake Superior. Since Broadwings (and most other hawks) are hydrophobic, and unwilling to fly over the big lake, they funnel down the North Shore and right over Hawk Ridge. Don't forget your binoculars!

Barred Owl

Strix varia

Large bark-colored owl

Wing span 4 feet

Strictly nocturnal

Call is loud booming "Who cooks for you . . . who cooks for you alllll"

One of the most unearthly sounds you are ever likely to hear may be the caterwauling of a pair of Barred Owls in a territory dispute. I remember well a September night on Duncan Lake. We were enveloped in darkness and silence when, from nearby, a low, booming "hoo hoo . . . hoo hoo" erupted from the black woods. Another owl responded from about 100 yards to the east. We stood in silence beneath the dome of stars, entranced. As the gap between the owls closed, the intensity of their territorial "hoots" seemed to grow. Then the night erupted in a cacophony of hoots, wails, and shrieks that sounded like a cross between the agonized screams of a tortured human and the cries of an injured Bobcat. It was an amazing experience.

The Barred Owl is more likely to be discovered by the ears than by the eyes in the Boundary Waters and Quetico lands. Nocturnal hunting of rodents in deep, dark woods and day-roosting in thick evergreens keep these owls from encountering many humans . . . even quiet, sensitive, nature-loving, granola-eating canoeists! But the Barred Owl cannot keep quiet.

On calm nights listen from the fringes of your campsite for the wondrous voice of *le chat-huant du nord,* the hooting cat of the north. The most vocal of North America's nocturnal raptors, the Barred Owl possesses an amazing array of hoots, wails, and screams, which it does not reserve solely for the breeding season but with which it will readily speak its mind in spring, summer, fall, or winter. Its song is a series of deep hoots often written as "who cooks for you . . . who cooks for you alllll" with the "all" coming out as a high-pitched, descending wail.

The Snowy Owl and the Great Horned Owl are the only North American owls heavier than the Barred Owl. Its 20-inch-long body unfolds to glide silently on wings 4 feet across, striking an unsuspecting Red-backed Vole with 1 pound of feathered furry. Its body is the color of spruce bark, with horizontal barring on the neck, vertically marked underparts, and rich brown eyes. It has no ear tufts or "horns" like the Great Horned Owl's.

Life in the dark

Barred Owls do about everything in the dark . . . except sleep. They are masters in their world of starry skies and branch-filtered moonlight. Fantastically equipped to hunt in low-light conditions, they feast on other active night-dwellers such as mice, voles, frogs, beetles, flying squirrels, grouse, and occasionally fish and crayfish.

Eyeballs as large as ours, with oversized pupils that contain an abundance of light-gathering rod cells, allow Barred Owls to see well in $1/100$ of the light that we would need to see equally well. Also, unlike most birds, their eyes are set in the front of the head so each eye's field of view overlaps, giving them binocular vision in a 70° arc, ten times more than nonraptor species. This comes in very handy for pinpointing prey. Their eyes are tubular and cannot be rotated in their sockets like ours, but fortunately their neck can rotate 270°.

Hearing, which is just as important as sight in ordinary nocturnal hunting, may be even more important on new-moon and overcast nights. Famous experiments with Barn Owls proved that in zero-light conditions owls use hearing exclusively and that they are not able to sense infrared or use vision under those circumstances. The facial disks around the eyes, which give owls their spectacled and hence "wise" appearance, serve as miniature radar dishes by collecting sound waves and focusing them on the ear holes located just below and off to the side of the eyes. One ear hole is slightly larger and lower on the face than the other, so sound reaches it at a different pitch and a fraction of a second later or sooner, depending on from which direction the sound comes. This enables the perched owl to triangulate the position of the potential victim within 1° to 2° in both the vertical and horizontal axes.

But sharp eyes and keen ears achieve nothing without a smooth and silent approach. Fortunately, Barred Owls have this covered, too. The outer primary feather, which forms the wing's leading edge, is finely serrated, forming a ragged edge that silences the noise of air rushing over it in flight. Most critters never know what hit them.

 Sparky says: Make your own "owl ears" by cupping your hands behind your ears and turning them into mini radar dishes like the owl's face disks. Now turn and focus them on some distant noise, such as a calling frog or singing bird. Take your hands away to hear the difference. It really works. Now try to call a Barred Owl into your campsite. When you hear one in the near distance, imitate its call, "who cooks for you," as best you can. Stop and listen for the owl's response using your owl ears. No giggling allowed.

You can see much on the way to the latrine. For those of you who have not yet visited the BWCAW, the government latrine, or "G.L." as it is affectionately known, is an outhouse without the surrounding house. The wood model well known to campers will soon be replaced by the "improved" fiberglass model. Since it has no surrounding walls, the path to the throne must be long enough to give users privacy from the campsite. Slow down and enjoy the walk, for here many people first encounter the Spruce Grouse, a bird so tame that humans have labeled it the "fool hen."

You may also see Spruce Grouse along the portage paths in Black Spruce–Balsam Fir–Jack Pine woods. Grouse often choose to ignore humans and go about their business of foraging in blueberry patches, nibbling on fungi, and snatching up insects.

The Ruffed Grouse and Spruce Grouse live side by side in much of the canoe country, so it is important to know how to separate these grouse cousins. Both chicken-size birds feed on the ground and also in trees but rarely fly long distances. The male Spruce Grouse has a black throat and chest and red "eyebrows," which are known as combs. The female is mottled brown. Neither sex has a crest like the Ruffed Grouse. Both sexes of Spruce Grouse have dark tails edged in chestnut; the Ruffed Grouse's rust or gray banded tail is bordered with a broad dark band.

Growing snowshoes

As the autumn days grow shorter, less light passes through the skulls of Spruce Grouse. The difference is detected by photoreceptive cells in the brain, which trigger hormonal release initiating the growth of "snowshoes" on their feet. The "snowshoes" are actually tiny scalelike growths that form a fringe along each toe, effectively doubling the foot's surface area. In a land of deep snows this adaptation is a lifesaver.

Spruce Grouse
Canachites canadensis

Chicken-size bird

Male has a black and white chest and red "eyebrows"

Tail is chocolate brown edged in chestnut

Female is completely mottled brown

Cutting it close

Spruce Grouse researcher Bill Robinson received a call one summer from a logger who had discovered a nest. The logger was felling a Jack Pine with a chainsaw when he noticed an incubating hen covered in sawdust, just 10 inches below his bar. The tree was spared, but clear-cut logging continued for acres around. The hen stood (sat) her ground. Miraculously, the clutch hatched successfully on June 27, 1966, and made the long trek to the surviving woods on the horizon.

Pass the needles, please

A bird known as the Spruce Grouse must eat spruce needles, right? Wrong. According to several researchers in Ontario and Upper Michigan, the winter diet is 99 percent Jack Pine needles. This diet causes many hunters to shun Spruce Grouse, claiming the flesh tastes like turpentine. This is true for birds shot from the winter through early spring, but what many don't realize is that come summer, grouse crave a diet of protein-rich insects, mushrooms, horsetail stems and tips, and moss spore cases, and more than once I have seen them blissfully eating blueberry leaves and berries in a large patch. Spruce Grouse eat few spruce, fir, or pine needles during the summer months. The result is that they make a fine meal during the early part of the grouse season.

All puffed up and nowhere to go

Sam and Seymour are two Spruce Grouse I came to know well at Seagull Lake, bordering the BWCAW. Both started their spring mating display just as the ice began to go out. In 1988 the show began on April 23, and in 1991 on April 22, with the ice leaving the main body of the lake less than a week later.

Sam's display area was at the intersection of two narrow foot paths on Dominion Island in an upland old-growth Jack Pine forest. He let me watch from less than 10 feet away as he fluffed out his black and white chest and neck feathers, fully erected his red eyebrows, drooped his wings, and completely fanned his tail, all while perched on a spruce branch 5 feet off the ground. He then flew down to the path throwing his body vertical just before landing. Strutting his stuff, he occasionally stopped and snapped shut his wide-fanned tail and then opened it again, producing a sound like a double deer snort or the sound made by blowing two quick bursts of air across your index finger. Drawing his body erect, Sam gave one or two stiff-winged flaps before flying up to another tree perch at head height. From there he flew to a mossy patch of ground, strutted, tail-snapped, stiff-wing flapped, and then repeated the whole performance, using the same four stations. I'd like to think that Sam's and Seymour's shows were all for me, but I believe that several very impressed hens were watching from the underbrush. The courtship displays of Spruce Grouse continue through May, with decreasing intensity as mating takes place and the eggs are laid.

 Sparky says: Try your hand at "ethology," the study of animal behavior in its natural environment. The Spruce Grouse, Sam and Seymour, provided me with excellent study subjects, but much of what I observed would have been long forgotten had I not written it down. Sit patiently with notebook in hand and take careful notes on the behavior of a cooperative critter. Campsite chipmunks and Red Squirrels make excellent subjects, as do stalking Great Blue Herons and birds at the nest.

Ruffed Grouse

Bonassa umbellus

Chicken-like bird with head crest

Tail is rusty or gray with a terminal black band

Male drums from downed log in spring and summer

Mama grouse, head down and wings slightly spread, charged straight down the portage. She was at ramming speed and heading directly towards me. It was a very convincing bluff. I had seen her six downy chicks scramble for cover before she went into attack mode so I understood her motive. Many June canoeists could relate a similar story. Portages seem to be the best place to see Ruffed Grouse during any snow-free season. Many hunters take advantage of this by combining a fall canoe trip with some "partridge" hunting.

Ruffed Grouse are rarely seen in flight and so are rarely seen by paddlers. As grouse expert Gordon Gullion says, "A grouse in flight is a grouse in trouble." Drumming males, though, can be heard from the water or the campsite, especially during the May mating season.

The Ruffed Grouse, a crested chicken-size bird, comes in two color phases: red and gray. This color shows most prominently on the tail, which in both phases has a terminal black band tipped in gray. In Minnesota, reds are more abundant in the south and grays more common in the north. Researchers believe grays to be better adapted to the cold. Studies in central British Columbia found that reds were more aggressive and better at securing drumming and nesting sites at higher densities than grays, but produced fewer offspring.

Grouse are mainly nonvocal birds, but males do "drum" to attract a female and defend their territory. Their drumming song sounds like an old International Harvester tractor starting up—"put, put, put, purrrr" and can be easily heard a quarter mile away. When they flap their cupped-wings, air vacuums form, which in turn fill with inrushing air to create the sounds. The drumming is such a low frequency, only 40 cycles per second, that Great Horned Owls cannot even hear it, much to the grouse's advantage. Many humans also have a hard time distinguishing the sound.

Ruffed Grouse most often drum from fallen logs. They drum from the same log year after year, rarely moving more than 600 feet from it during the spring. One large White Pine log, cut around 1900 in the Cloquet State Forest, has been in use by a steady stream of males since 1931.

The aspen-grouse connection

Ruffed Grouse and Quaking Aspen—it is no coincidence that their North American ranges neatly overlap. Grouse use aspen in every season and stage of life. In spring, males drum beneath an aspen canopy, often from the top of a fallen aspen log. Once mated, the hen selects her nest location carefully. The conscientious grouse mother finds an ideal spot to be at the base of a mature aspen with a 50-to-60-foot clear view all around. Not only are predators easy to spot but she can also hop off the nest and up into the canopy for a quick meal of leaves while still keeping an eye on her eggs. The precocious young need dense stands of aspen suckers for protection. In late June or early July adult grouse become florivorous, feeding nearly exclusively on the flower buds of male Quaking Aspens. Ruffed Grouse continue to live on this diet until they switch to the flowers themselves.

Quaking Aspen flower buds are high in nutrients such as minerals, protein, and fat. Balsam Poplar buds may be just as nutritious, but grouse reject them because of the gooey resin that coats the bud scales. A study by ecologists Tester and Huempfner on the winter diet of Ruffed Grouse at Minnesota's Cedar Creek Natural History Area found that 87 percent of their diet was Quaking Aspen buds, with the buds of Big-tooth Aspen, Paper Birch, and Black Cherry making up the balance.

You've probably seen grouse loading up on buds for the evening, silhouetted in the tops of aspen trees against the darkening fall or winter sky. It takes only fifteen to twenty minutes of feeding at dusk to fill their crops with 3 to 4 ounces of buds, the equivalent of a 150-pound human eating 27 pounds of food at one sitting.

During high cycles of the grouse population, Quaking Aspens really take a beating. When the trees are stripped of

Snowdiving

Ruffed Grouse sleep under the snow in winter. After plunging in, they burrow for up to 20 feet (but usually only 3 feet) to confuse predators. Air temperatures may be -40° F just above the snow, but burrows under two feet of snow rarely fall below +20° F. A cold, low-snow year is bad news for Ruffed Grouse. With no protection from the cold, many expend too much energy trying to keep warm and die.

their flowers, reproduction nearly halts. What's an aspen to do? Fight back! Grouse researcher Gordon Gullion had the Quaking Aspen flower buds analyzed after noting that grouse were ignoring their usual favorite food one winter. Testing revealed that the aspens had exuded onto the bud scales a resin that contained a phenol compound, making the buds very difficult for grouse to digest. Can the aspens "talk," warning one another of impending grouse invasions by the use of wind-borne pheromones? That question remains to be answered, but one thing is sure: Ruffed Grouse and Quaking Aspen are intimately linked within the complex ecosystem of the North Woods.

 Sparky says: Listen for a grouse drumming nearby. Try to imitate the sound by beating on your chest with your fists. Get closer. Can you get close enough to see him? This is a tough challenge, but patience and perseverance will reap great rewards.

Common Nighthawk
Chordeiles minor

Dusk fast approaches. You sit on the quartz-eye Saganaga granite soaking in the setting sun and savoring your meal of Lake Trout (or texturized vegetable protein meat-substitute chili). Suddenly from overhead a nasal "peent" breaks the silence. Looking up, you see several Common Nighthawks performing tactical aerial maneuvers. Not for mere pleasure, you realize, but for survival. They twist, turn, and dive with mouths agape in pursuit of flying insects . . . including the mosquito. The Common Nighthawk becomes your friend—partner in the elimination of the winged vampire.

The Common Nighthawk, a 10-inch-long cryptically-colored bird, is nearly invisible when it sits on a slab of bare lichen-splattered rock or on a tree branch. Its bill appears tiny. The bird transforms when on the wing, becoming a graceful flier floating on foot-long wings, each boldly marked underneath by a white stripe at the bend of the wing. This infallible field mark is visible from long distances. And remember that tiny bill? In flight it opens to reveal a huge gaping mouth that sucks in insects effectively.

Narrow, pointed, foot-long wings

White stripe under each wing

Call is nasal "peeent"

Active at dusk

Aerial vacuum

The Common Nighthawk, neither a hawk nor strictly nocturnal, belongs to the family Caprimulgidae, which includes the familiar Whip-poor-will. It does, however, "hawk" for insects in flight by catching them while on the wing.

Pointed, narrow wings provide excellent aerial maneuverability. Numerous elongated rod cells in the eyes aid in light gathering for dawn and dusk insect hunting. But most importantly, the Common Nighthawk has a huge gaping mouth to catch any edible airborne critter. One Common Nighthawk took in 500 mosquitoes in one evening. Another went on a feeding frenzy, inhaling 2175 flying ants.

Anyone have a comb?

The Common Nighthawk has a comb, not a hair comb, nor a comb on top of the head like a rooster's, but rather a "feather comb" or "louse comb." The comb is the "toothed" middle claw of the foot. The bird may run the comb through its feathers to rid itself of parasites.

Shepherds beware!

Members of the family Caprimulgidae are commonly known as "goatsuckers." The myth that gave rise to this ridiculous name claims that birds such as the Common Nighthawk, Whip-poor-will, Chuck-will's-widow, and Common Poorwill would swoop down on innocent herds of goats, clamp their gaping mouths over the udders, and suck the goats dry of milk. Of unclear origin, the myth was known in Aristotle's time.

The gap between myth and reality may be just a short extrapolatory leap. A shepherd may have seen Nightjars (a Eurasian relative of the Common Nighthawk), attracted by insects kicked up by a herd of goats, swarming about his animals. When, just by chance, the goats gave milk poorly the next morning, the logical but incorrect assumption was made.

Lunarphilia and the nesting cycle

Common Nighthawks and their cousins, the Whip-poor-wills, are moon-loving species, or lunarphilics. Studies have shown an increase in feeding flights on full moon nights for both species. A. M. Mills discovered that Whip-poor-wills time the laying of their eggs to the lunar cycle. Eggs usually hatch during a young waxing moon so that during the critical first two weeks of life the young receive an ample food supply from parents who utilize the increasing moonlight for profitable hunting. The next waxing moon leading to a full moon occurs at another critical phase in the life cycle, that of fledging. Common Nighthawks exhibit this same lunar coordination but to a lesser extent, probably due to their greater dependence on dawn and dusk feeding and limited nocturnal activity.

Lawn chairs and lazy August evenings

The spectacular late summer southbound migration of the Common Nighthawk is a sight to behold. But one must be in the right place at the right time with the right frame of mind to catch it. The last time I witnessed it was August 27, 1989. Driving down the North Shore, I noticed that I wasn't the only one southbound. I rushed to my home on the central hillside of Duluth, set up my lawn chair on the roof, pointed my binoculars lakeward, and started counting. In one hour I counted 3569 Common Nighthawks streaming over downtown Duluth on their long journey to their wintering grounds throughout South America.

This pales, though, in comparison to Mike Hendrickson's spectacular count of 43,690 Common Nighthawks seen between 5:35 P.M. and 8:20 P.M. on August 26, 1990, with a mind-numbing 23,000 of these coming in just one hour. The entire count was done while sitting in a lawn chair on Lake Superior's North Shore.

Lake Superior acts as a funnel to autumn migrating birds. All the Common Nighthawks breeding in northern Minnesota and northwestern Ontario head south in mid-August. In their southbound movement they eventually hit Lake Superior, which to an insect-eating bird is a virtual desert, a place to be avoided at all costs. They funnel down the shore on calm, clear, and warm evenings, with inland birds joining their ranks until, on rare occasions, they become a huge mass moving through Duluth, which is the end of the funnel.

 **Sparky says:** Practice the skills that prey animals use, such as camouflage, by which nighthawks blend masterfully into their surroundings when at rest. "Predator-prey" is a game that beautifully illustrates this principle. One person is chosen to be the predator (wolf, warbler, Whirligig Beetle, Wolverine, etc.); the rest of the players are prey (let them choose). The predator closes its eyes for sixty seconds, which he or she counts out loud . . . very loud. Meanwhile, the prey hide, using all the camouflage secrets (coloration, stillness,

quietness, concealment, etc.). The only catch is that wherever they hide they must at all times be able to see the whites of the predator's eyes.

Opening his or her eyes, the predator scans the woods without moving from the spot, though he or she is allowed to bend down. When the predator spots a prey, the prey is called out of hiding. After five minutes, all the captured prey become predators, clustering together and counting to sixty while the remaining prey move to a new spot at least 10 feet closer. The last prey left unfound becomes the next game's predator.

Discuss what camouflage strategies were used and which worked best. How do the wildlife you've seen on your trip escape detection? What senses do predators use to locate prey?

The loud, raucous rattle of a flying Belted Kingfisher, by which most northern canoeists first encounter the "king of fishers," demands attention. A bird of shores, wherever water meets woods, the kingfisher flies a regular beat, stopping only to perch and fish. Paddling a shoreline, you may chase the kingfisher from perch to perch in front of you until it hits the edge of its territory and doubles back. The shoreline territory may be as short as 500 yards or as long as 5000, depending on availability of small fish and crayfish. The kingfisher hunts just like its fellow piscivore, the Osprey, perching, hovering, and plunging to catch fish. Watch as it intently eyes the water from its perch for any fishy movements. To get into striking position the kingfisher usually leaves the perch and hovers, and at the critical moment plunges beak-first into the water, quickly returning to a perch to eat its catch or regroup for another attempt.

The Belted Kingfisher is an anomaly in the bird world. In most species the male wears the gaudy colors, but the female kingfisher reverses the roles by sporting a wide rusty "belt," which the male lacks. This is doubly strange since among their little cousins, the Green Kingfishers of south Texas, the male wears the rusty belt, which his mate lacks.

The Belted Kingfisher is slightly larger than a robin, with a head that looks too big for its body because of a hefty daggerlike bill and a shaggy crest. Both sexes have a blue head, back, and breast band on a background of white.

Kingfisher shades

Day-active birds possess trichromatic color vision like our own. Colored oil droplets in the cone cells of the retina, which allow for such vision, occur in differing percentages in different birds. Hawks and most songbirds have 10 to 20 percent red oil droplets, which makes their color vision probably like ours. Kingfishers have 60 percent red oil droplets, which reduces the glare on water and allows them keen underwater vision.

Belted Kingfisher
Ceryle alcyon

Robust 12 inch long bird

Stout daggerlike bill and shaggy crest

Male is blue except for white belly and neck ring

Female wears a rusty "belt" around her belly

Call is a raucous rattle

The tunnel of love

Belted Kingfishers nest in burrows dug into eroding banks. The kingfisher repeatedly flies beak-first into the bank until a ledge is formed. With its bill as a pick and its feet as a shovel, it flings dirt out and back. The male and female take turns digging during the one to three weeks it takes to complete the excavation of the tunnel and football-size nest chamber. The narrow tunnel is usually from 3 to 6 feet long but may sometimes even reach 15 feet.

Six or seven eggs are laid. Both sexes incubate the eggs during the day, changing positions only when they recognize their mate's call outside the burrow. The female stays on the eggs all night while the male sleeps in a roosting burrow he has dug nearby.

Newly hatched young are fed regurgitated fish. Parents are kept busy, since each nestling requires an average of 11.2 fish per day for maximum growth. So, for a half dozen hungry nestlings, each parent must catch 2.6 fish per hour for the entire fifteen hours of daylight, day in and day out.

To teach the nestlings how to fish, the parents stun small fish and float them past the perched babies. The young don't take to water instinctively or immediately. Eventually they understand that filling the stomach means a wet head. A mere ten days after fledging, the young are chased out of their parents' territory to find their own fishing holes.

Alcyon

Alcyon, from which comes the kingfisher's Latin name, was a mythical Greek woman who so mourned her shipwrecked and drowned husband that the gods took pity on her and turned them both into kingfishers, to enjoy the watery realm together forever.

Pike snacks

Northern Pike have been known to eat plunging kingfishers on occasion, a feat that must require incredible timing.

Sparky says: Imagine yourself a fishing kingfisher. As you hover 20 feet above the water, your eyes search for darting minnows and fingerlings just below the surface. A big fat minnow comes close and you plunge head first, eyes locked on your prey, grabbing it below the water with your mouth. You fly up to a perch, swallowing the fish whole.

Wherever bark beetles live in the coniferous forest, Three-toed Woodpeckers and Black-backed Woodpeckers quickly move into the neighborhood. Bark beetles, wood-boring beetles, and beetle larva are the preferred cuisine for these woodpecker cousins. Their search for beetle morsels will give you your first clue to their presence. The sound of flaking bark in older stands of spruce, fir, and Jack Pine, bogs, or newly burned areas announces the Three-toed and Black-backed Woodpeckers' food foraging. A canoeist most likely will encounter these birds while portaging, gathering firewood, strolling to the latrine, or picking blueberries in a recent burn.

The two woodpeckers have in common a three-toed foot. This is a difficult trait to see in the field (even for experienced bird-toe counters), so we must rely on more obvious traits.

From a distance, both species appear to be 9 inch-long, black-and-white, woodpecker-shaped birds. Fortunately, like many boreal birds, they are very trusting and allow humans a close approach. Upon inspection you'll see that, yes, the Black-backed does indeed have a solid black back, while the Three-toed has a band of black and white barring running down the middle of its back. If you are fortunate enough to see both species together on one tree trunk you will notice that the Black-backed appears larger and "neater," with well-defined color patches and patterns. The Three-toed appears slightly bedraggled. The males of both species sport a jaunty yellow cap, which the females lack.

The Three-toed calls with a sharp "kik." The Black-backed gives more of a "krik" call and, when flying, a grating staccato growl. Both calls differ from the more familiar calls of the Downy and Hairy Woodpeckers, who give softer-sounding "pik" or "peek" calls.

Only three toes?

Most woodpeckers are zygodactyl, with two toes pointing forward and two toes pointing backward. The Three-toed and

Black-backed Woodpecker
Picoides arcticus

Three-toed Woodpecker
Picoides tridactylus

Length 9 inches long

Males have a yellow "cap"

The blackbacked's back is completely black

The three-toed's back is barred black and white

Black-backed are tridactyl, with only one toe pointing back. But having one less toe than other woodpeckers appears not to hinder the birds in the slightest. Very stiff tail feathers act as a third bracing foot, allowing the woodpecker very solid support on the tree. The secure points of contact enable it to "hitch" up and down vertical trunks.

Fire as friend

Fire is a necessary component of any wilderness ecosystem. Fire burns up the dead vegetation, renews the soil by releasing nitrogen from organic matter, and creates new openings with abundant and easily accessible food for a variety of critters.

In July 1988, I revisited the site of the Clearwater fire, a 440-acre fire that crowned out and for two days in May burned hot and hard just south of the BWCAW's Daniel's Lake. Hotshot fire crews, Canadian water bombers, and torrential rains finally doused the inferno. On my hike through the two-month-old burn, already regrown with waist-high vegetation, and alive with grasshoppers, and other insect life, I found many birds uncommon to the mature woods of the Boundary Waters–Quetico, including Eastern Bluebirds, Barn Swallows, and American Kestrels. The newly created wide open spaces and abundant food attracted these birds.

Also in the burn were three Black-backed Woodpeckers, no doubt attracted to the fire-damaged trees by the abundance of wood-boring beetles. Researcher James Koplan recorded a 50-fold increase in Hairy, Downy, and Three-toed Woodpeckers after a 10-acre burn in a Colorado coniferous forest.

He found the Three-toeds to be very specialized feeders, uncovering wood-boring beetles by vigorously flaking bark with a sideways bill movement, working nearly exclusively on the trunks of spruces. They can be extremely efficient—James Koplin and Paul Baldwin recorded Engelmann Spruce Beetle populations being reduced by 45 to 98 percent.

Home life

The Black-backed and Three-toed have nearly identical ranges in North America. Both occupy roughly the range of their nearly exclusive feeding tree, the spruce. The Black-backed is slightly more common in the east, with the Three-toed extending farther south in the Rocky Mountains. Interestingly, the Three-toed is the only North American woodpecker also found in Eurasia. The first Minnesota breeding record was the June 26, 1981, discovery of a nest adjacent to the BWCAW in Cook County, just off the Gunflint Trail.

The cousin woodpeckers have similar nesting behaviors. Both the male and the female excavate the nest cavity. All four nests I've located in the BWCAW have been 10 to 40 feet up in live Jack Pines with presumably dead hearts. The hole was about 2 inches in diameter with a "doorstep," a beveled lower lip. The trunk immediately around the hole had been totally debarked.

In early May, the female lays a clutch of four white eggs (no need to lay camouflaged eggs, since they are protected from any predator's view in the cavity). Both male and female brood the eggs and feed the young when they hatch after about two weeks of incubation.

Early one morning I watched and recorded the activities of a pair of Black-backeds and their nest cavity of hungry nestlings just to see how hectic life can get for Mom and Dad.

6:19   Female feeds young. From inside cavity comes constant "chipping" of 2, 3, or 4 young.
6:22   Female returns with food and feeds young.
6:24   Male lands on nearby tree.
6:25   Female goes to nest hole with food. Male follows close behind. Two young visible, peeking heads out hole.
6:28   Female returns with food and feeds constantly squawking young.
6:33   Male lands at nearby tree and calls. From there he flies to nest hole with mouth full of goodies.
6:36   Male returns. He must have a nearby abundant source of grubs or beetles to make such frequent feeding trips. When done, he sidles

around to the back of the Jack Pine before flying off. He calls.

6:39 Female feeds the noisy nestlings.
6:42 Male feeds young. I see him foraging nearby.
6:44 Male feeds young.
6:45 Female shoves more insect-delight down the little gaping bills. Male flies up and takes over as female flies off to reload.
6:48 Male feeds young.
6:50 Male feeds young.
6:53 Male feeds young.
6:55 Male feeds young. Male on a furious pace. Can he keep it up?
6:56 Relief! Female arrives with the groceries.
6:58 Female feeds young.
7:09 Female feeds young. Now she's drumming repeatedly. Why?
7:17 Male comes to nest hole, circles trunk, and pops into the nest cavity. Is he removing young's fecal sacs?
7:22 Male leaves nest cavity.
7:32 Male feeds young.
7:39 Male feeds young.
7:45 I'm cramped and wet. Time to leave.

I returned the next day to a very quiet nest tree. Evidently the little hellions had flown the coop and were now testing their wings, fluttering awkwardly from trunk to trunk. But the parents get no rest; the newly fledged young still squawk to be fed. They gradually gain the skills to feed themselves, and in less than a year they might be frantically scrambling to fill 2, 3, or 4 wildly gaping little bills of their own. What goes around comes around.

 Sparky says: Imitate the food-gathering behavior of the Black-backed and Three-toed Woodpeckers. Gently flake the bark from a fallen spruce or Jack Pine, looking closely for insects hidden beneath. Debark the log all the way down to the wood. Do you see any bore holes made by the juicy larvae of bark beetles?

Athapaskan legend

One legend to come out of Alaska, most likely from the Athapaskan peoples, portrays the male Three-toed as a very jealous man. The elders told of a time long ago when starvation and great famine spread throughout the land. The desperate woodpecker-man killed and ate his own spouse. After the cannibalistic murder, he cleaned her greasy fat from his claws by rubbing them atop his head. Ever since, the Three-toed has been branded with the yellow cap for all to know of his evil deed.

Pileated Woodpecker
Dryocopus pileatus

Paddling a placid lake, you see a large black bird flapping from an island to the mainland, looking for all the world like a plain old crow until you notice the large flashing patches of white on the wing. Then you know that you've seen North America's largest woodpecker, the Pileated. Pileated, which means "capped," refers to its red crest; it can be pronounced "pill-ee-ated" or "pie-lee-ated." The bird doesn't care.

The Pileated has always appeared to me to be a prehistoric bird. I can vividly envision it winging its way through some giant fern forest of the Pleistocene. Its size and look are that of an archaic, long extinct, flying fossil better suited to some ancient epoch. I am always in awe of them.

You don't really get a feel for the hugeness of this crow-sized woodpecker until a Pileated flies right in front of you in the understory of the big forest. Powerful wings provide audible testimony to the massive amounts of air they move with each flap; then with a "whoomp" it lands vertically on the trunk of a large Jack Pine. You don't get a feel for the power of the Pileated until it rears back its head and with jackhammer blows sends the wood flying. Chips ranging from match-size to match box-size are excavated from the living tree so the bird can reach the burrows of the tart-tasting Black Carpenter Ants. Rectangular holes in the exposed roots and lower trunks of large trees, along with "sawdust" mounds at the base, are mute signs of the workings of the "king of woodpeckers."

The Pileated is unmistakable. It is 17 inches from beak to tail, with a wingspan of 2 feet. Both male and female have a flame red crest, white chin, large white underwing patches, and a white stripe running from the beak below the eye and down the neck. The male additionally sports a red moustache.

Their forest-piercing "kwuck-kwuck-kwuck" comes sometimes in a steady staccato and other times with a rise and fall in volume. Either way, the ringing call is audible at several hundred yards.

Large crow-sized black woodpecker

Flame-red crest and white face stripes

Large white patches on underside of wings

Call is "kwuck-kwuck-wkuck"

Lean, mean, food-gathering machine

Large Black Carpenter Ants constitute the bulk of the Pileated's winter and summer diet in the North Woods. One Pileated was collected with 2600 Black Carpenter Ants in its stomach. Now that's a feast! The carpenter ant, as its name indicates, works in wood. It invades outwardly healthy-appearing trees by excavating burrows in the softer and invisible heart rot. The invasion often starts in exposed roots, and the ants eventually work their way up the tree.

Anybody for a Sweet Tart?

How Pileateds recognize a carpenter ant–infested tree is not well known. Most authorities believe woodpeckers hear their prey insects inside trees. It has been proposed, though, that Pileated Woodpeckers can smell the formic acid carpenter ants excrete as body waste. The tangy formic acid is a substance used in making the popular candy Sweet Tarts.

I need an aspirin

Once a Pileated has found that ants are present, the real work begins. Anchored securely to the tree by feet with two sharp-clawed toes pointing forward and two pointing back and propped by twelve stiffened and pointed tail feathers, the Pileated begins chiseling away wood with fierce blows, its hard-edged bill strengthened with structural ridges. It uses its bill as a hammer, chisel, and pry bar.

Imagine pounding your nose into trees all day to make a living. Imagine the horrible headaches. How does the Pileated prevent brain damage and concussion? Ornithologists have discovered several amazing adaptations that allow woodpeckers to take a pounding and keep on pecking. The bill is anchored at the skull by a very thick honeycombed bone that acts as a shock absorber. Strong neck muscles that can propel the head

Homeless

Calamity struck a Pileated pair in Everglades National Park in Florida. A wind storm snapped their nest tree in half at the exact point of the nest. With the "roof" gone and the eggs exposed, the female took action. She moved the eggs to a new cavity by flying with one egg at a time in her bill. Egg carrying has rarely been documented in birds, but Frederick Kent Truslow photographed this event on April 16, 1966; the amazing photograph can be seen in the May 1970 *National Geographic* magazine.

forward at 15 miles per hour can also help to diffuse the jarring force of the blow.

A liquid membrane between the skull and brain case prevents the brain from being severely jostled and damaged by repeated hammering. Rumor has it that this inspired the invention of the water-packed football helmet, which reduced head injuries significantly. (Today the water-packed helmet has been replaced by the lighter air-cushioned helmet.)

Fishing for ants

Finally the bird reaches the ant burrows. Now the tongue must bring home the booty. Contrary to popular belief a woodpecker actually has a short tongue. But the tongue is connected to a long, branched hyoid bone-muscle apparatus that wraps around the skull and actually is anchored in the nostril of the upper mandible. The tip of the tongue is horny and armed with many bristles that act as little "ant hooks" to snag them from their burrows. A sticky saliva coating helps hold the ants. The Pileated extends its tongue/hyoid down an ant burrow, feeling in every nook and cranny with the tip. Laden with touch sensitive cells called Herbst corpuscles, the tip can identify a dormant carpenter ant. The ant is snagged with the tongue tip bristles and held with sticky saliva. In a split second, the tongue/hyoid recoils, bringing the ant to the Pileated's mouth. Amazing!

A solar home

Pileated nest cavities usually face east or south, presumably to catch the morning light and the resulting warmth of a southern exposure.

Sparky says: Try a sample of woodpecker cuisine. Pileated Woodpeckers enjoy tart-tasting ants, so why shouldn't we? Most ants are edible. Pinch the head off a live ant and just pop the rest in your mouth. It's an incredibly tangy treat. Formic acid, the same stuff found in Sweet Tarts, provides the sweet flavor zing. Compare and contrast the flavors of several ant species.

Gray Jay

Perisoreus canadensis

Robin-sized

Appears soft and fluffy

Gray back, white breast

Common campsite visitor

The Gray Jay is most likely the first bird you will meet at a Boundary Waters–Quetico campsite. Before you can say "Lake Saganagons," a Gray Jay family "floats" from spruce to fir and into your campsite to check out the new neighbors and the goodies they brought. This bold and brash jay has made a reputation (and a nickname) for itself by its practice of snatching any and all food left out at a campsite. They have even been known to land on the edge of a hot frying pan to grab a piece of that nearly perfect Bisquick pancake a hungry camper was just about to flip! These habits have earned them the ungracious nickname of "camp robber." Stupid, tame, friendly, dumb, brave, bold, brash, and brazen are all adjectives used to describe the Gray Jay. But no canoeist who leaves the North Country will ever forget the antics of the Gray Jay.

The Gray Jay's overall gray (the back is gray while the nape and crown are dark gray) is brightened by a white breast, belly, neck-collar, and forehead. Immature birds are sooty-gray from beak to tail. Slightly larger than an American Robin at 11 inches long, all Gray Jays appear fluffy because of their very long down feathers, which may be 2 1/2 inches in length. Such insulating feathers offer essential protection to a bird that is a year-round boreal resident.

Like most of the birds in the family Corvidae, which includes the jays, crows, and ravens, the Gray Jay has many songs and calls in its repertoire, including a high whistled "wheeoo," a harsh "chack," and a high-pitched chatter that imitates a Merlin beautifully.

The world's largest saliva glands . . . or close

You will soon learn that the camp robber's appetite seems unquenchable. This inquisitive bird with friendly eyes takes and takes and takes, flying off with its booty only to return minutes later for more. Oddly enough, this behavior directly relates to the fact that Gray Jays have some of the largest salivary glands of any bird. According to ornithologist Douglas

Dow, this adaptation allows the jay to take excess food, mix it with sticky saliva, and make marble-size food pellets that can be stored for winter use in branch crotches, under loose bark, or in tufts of pine needles. Ample glutinous saliva holds the food pellet in place and may even act as a preservative.

These carefully stored food pellets provide energy-rich insurance against starvation at a time of high energy demands. Not only does the Gray Jay need to stay warm in winter but it also builds a nest and starts a family then. That takes energy! Construction usually begins in late February, with jays sitting on eggs by mid-March, when the snow still lies waist deep in the woods and temperatures can plummet to -20° F.

The nest is a super-insulated home. Tucked into the dense branches of a spruce or fir to reduce radiant heat loss at night, the nest is usually on the south side of the tree to catch the winter sun's warmth. "Bulky" well describes the nest, which has a framework of sticks layered with shredded cedar bark, cast-off cocoons, and wasp nest pieces. Options for lining include fine grass, Snowshoe Hare hair, Moose hair, deer hair, grouse feathers, and the jay's own down. Now that sounds warm and comfy to me! Deer hair, an excellent insulator because it is hollow, is especially prevalent in Gray Jay nests.

Where do Gray Jays find deer or Moose hair? Since they are scavengers on nature's cleanup crew, they are are always looking for deer and Moose carcasses and hence food. And since nothing in nature is wasted, the Gray Jay, after eating, may fly off with a beak full of hair to line its nest.

 **Sparky says:** Try to feed a Gray Jay out of your outstretched hand. Secretive during early summer, these birds literally mob your campsite for a handout come late July through fall. Bread, biscuits, bacon, pancakes, and Rye Krisp are all good bait. Watch them as they fly off with their treasure to see if it is cached or eaten.

Make mine a whiskey

These aggressive yet friendly birds are also known by the vernacular name of "Whiskey Jack." This may be a corruption of the Ojibwa Indian name of Wiskaju. But if you want a colorful North Woods explanation involving a beautiful Ojibwa woman, a lumber-camp ax murder, and drunk birds, read "Four-bottle McGovern and the Whiskey Jacks," related by Bill Magie in the book A Wonderful Country. The true source of this nickname may be lost forever in the morning mists of history.

A bar of soap to go, please

Gray Jays will eat anything dead. But since their bills are very weak, they only rarely kill rodents themselves. Ants, ant eggs, berries, spiders, oatmeal, beans, and pancakes make up just a partial list of their known diet. Less understandable are observations of Gray Jays picking at bars of soap and candles.

Common Raven

Corvus corax

Length 2 feet

Jet black

Massive beak

Wedge-shaped tail

Low guttural croak

The ubiquitous Common Raven—master of the northern sky, lord of the hinterlands, the all-knowing and all-seeing great croaker, the black one, noble fool and trickster to Native Americans, messenger sent forth by Noah from the ark, and whisperer of the earth's secrets to the Nordic god Odin—it's anything but "common."

In 1936, when T. S. Roberts wrote his epic *The Birds of Minnesota,* the Common Raven was nearly extinct from the state due to the extensive logging of the old-growth forests. Fortunately the big trees are back and the raven soars over the canoe country again. Paddlers may see several ravens in a day, some performing aerial acrobatics to impress a mate or simply to have fun.

A member of the order Passeriformes, the Common Raven is the largest songbird in North America. To me, a 2-foot-long black bird with a bill capable of cracking a seal's skull is not a songbird, but their guttural croak, which is much lower pitched and throaty than that of the American Crow, serves well as their song. The croak is the music of wildness. I truly believe that ravens have an extensive language all their own, by which they communicate many things with each other.

Ravens have a 4-foot wingspan and weigh four times more than crows—a good way to tell them apart if you're holding them in your hands, but meaningless when they're flying overhead. Crows have a three-sided squared tail, while the raven's tail is four-sided and wedge-shaped—a trait most visible on soaring birds. Crows rarely soar. Some claim that ravens fly with their mouths open, a fact I have verified on several occasions. Check it out for yourself.

As I write this, one lone ravenette squawks from its nest high up in a 130-year-old Jack Pine just outside the cabin door. Its mother, in a neighboring tree, is trying to coax it to flight. I found a fledged youngster on the ground yesterday, obviously flustered by life out of the nest. Young of this age have a brown head and breast with the most incredible milky blue eyes.

The most intelligent bird

That Common Ravens could speak German came as quite a surprise to me. In the winter of 1988, while traveling through Europe, I stopped in at the world-renowned Alpenzoo near Innsbruck, Austria. A raven stuck a stick through the wire of his spacious cage; I grabbed it and began a serious tug-of-war that lasted several minutes. I let him win and started to walk away. But he was not through, and let me know it by hopping ahead of me and sticking the same stick through the cage wires. I played for awhile but soon tired of the game and walked on. "Achtung, achtung," a hoarse voice demanded from behind. I wheeled about, completely at attention, expecting a zoo official, but instead looked square into the intelligent eyes of my raven friend. I've never taken a raven for granted since.

Raven intelligence is well known. It allows them to count, play, remember, and learn by watching others. Their creativity shows in their food-gathering methods in the high Arctic.

Teams of seven or more Common Ravens have been known to kill young caribou by taking turns pecking at the caribou's eyes until it falls down and can be given the coup de grace. Similarly, two ravens can dispatch a seal pup. One covers the exit hole in the ice while another drops down and cracks the seal's skull. Town ravens in Yellowknife and Inuvik dine in a more civilized manner by knocking over garbage cans or lifting off loose lids to get at the pickings.

Researcher O. Kohler learned that ravens can count. He trained them to open the box with the same number of spots on the lid as on a key card. This card, with up to six spots on it, was laid in front of a raven, who then would open the appropriate box to receive its reward. Kohler changed the size, shape, and arrangement of the box spots while the cards remained unchanged, and even reordered the boxes, all to confuse the raven. But the noble beast passed every test with flying colors, proving that ravens can count . . . at least up to six!

Play is often attributed only to "higher" animals, but ravens

A full pantry

Ravens require a high-protein diet, unlike their crow cousins, who can live on grain. To these ends Common Ravens will kill and cache Red Squirrels and Snowshoe Hares for use during the lean times.

Fear tactics

Great Horned Owls are the only predator of the raven. Strangely, ravens seem unwilling to mob or attack owls as most birds do.

Far-ranging ravens

The Common Raven ranges from the mountains of Guatemala to the Arctic Circle in North America. It is found throughout Eurasia, in Africa's northern Sahara Desert, and on the slopes of the Himalayas.

seem to love a good game. David Bruggers, who studied Common Raven ecology in Minnesota's Superior National Forest, watched them "count coup" on feeding Bald Eagles. One would lean in toward the preoccupied eagle, quickly pull its tail, and jump away to rejoin the watching ravens, who no doubt were enjoying a good laugh. Flying ravens have also been observed playing "chicken" with the ground, seeing who could come closest to it before pulling out of a dive. Even a tattered piece of deer hide becomes a toy in the beak of a raven, who tosses it to and fro.

Ecologist Bernd Heinrich believes that ravens "use" more dangerous predators, such as wolves and humans, for their own benefit. He cites the example of Miles Martin, an Alaska trapper who was in the bush running his trap line when a Common Raven repeatedly flew over his shoulder from behind. As it flew by, the raven would look back at him, as if to say "C'mon, let's go." (This well-known raven behavior is known as the "follow-me" display.) Miles went to the spot that the raven had "marked" with repeated dives, and there discovered a sleeping Moose. Coincidence? Or an intelligent critter equating gun-toting humans with Moose entrails?

Sparky says: Make up your own raven legend like the traditional ones. For example, the Norse god Odin carried on his shoulders two ravens, Munin (memory) and Hugin (thought). By day he sent them to journey to all the corners of the earth. Returning by nightfall, the ravens whispered the dark secrets of the world into Odin's ears. With this universal knowledge, Odin counseled other Nordic gods.

Let your creative force loose. Write your legend down. Embellish it. Have your group share their legends around the evening campfire.

"Chickadeedeedee. Chickadeedeedee." The Black-capped Chickadee calls its own name. Especially common in late summer after the nesting season, this call gathers a foraging flock to reassemble.

A high whistled "feeeebeee," with second syllable lower, is considered the Blackcap's true song. It can be heard any time of year but more frequently in spring during the mating season.

A black cap and black bib separated by snow white cheek feathers distinguish this tiny 5-inch ball of gray fluff from its close cousin the Boreal Chickadee. The Boreal Chickadee, of the same size, appears dirtier because of its brown cap, grayer cheeks, and darker flanks. Both sport the black bib.

"Zhicka-zhnee. Zhicka-zhnee." The Boreal sounds like a Blackcap with a bad head cold. Their distinctive nasal call is outnumbered by the Blackcap's "chickadeedeedee" in most areas of the canoe country.

I would diagnose both species as chronically hyperactive. Just as you get one chickadee located in your binoculars, it will hop one spruce branch over and out of view. Known as "power flitting," this is the chickadee's food finding techique. It gleans spider eggs, plant lice, bark beetles, moth eggs, and caterpillars from the bark and branches of trees.

Chickadees are readily visible at most campsites and can be "pished" in for a closer look. With clenched teeth, loudly repeat "pshhhhhhh" over and over. Pause. Any chickadees, warblers, or nuthatches in the area will likely fly in for a better look at the strange animal making the strange noise.

Small balls of fire

To maintain their normal body temperature of 108° F on a below-zero winter day, chickadees must eat nearly nonstop from sunup to sundown. A wintering Blackcap or Boreal Chickadee must spend twenty times as long feeding in -8° F weather as on a warm spring day. But flitting about gathering this life-saving food can raise a chickadee's heart rate to one

Black-capped Chickadee
Parus atricapillus

Boreal Chickadee
Parus hudsonicus

Tiny round fluffballs of birds
Length 5 inches

Black-capped has a black cap and bib, and white cheeks

Boreal has a brown cap and "dirtier" sides and face

The Boreal's call is "zhicka-zhnee, zhicka-zhnee"

"HISSSS!!!"

Female Black-capped Chickadees make an explosive hiss when disturbed while sitting on eggs in their dark tree-cavity nest. If that doesn't scare the wits out of a potential predator, I don't know what will.

Home construction

Woodpeckers aren't the only home excavators around. Both species of chickadees carve out their own nest hole in soft, rotting tree stumps. To keep its address anonymous, the little home builder flies off with beakful after beakful of wood chips and drops them in an out-of-the-way place.

thousand beats per minute, twice the resting rate, thereby eating up precious calories and turning foraging itself into a life-threatening situation. Large, energy-rich, and easily gathered sunflower seeds at feeders are a boon to winter chickadees. Less time and effort are spent in procuring the seeds, and the caloric gain is greater. A steady diet of sunflower seeds can raise a chickadee's fat deposits by 4 percent.

Chickadees cache a percentage of the seeds taken from the feeder for lean times. Colorado State University researchers have shown that Blackcaps have a high recovery rate for seeds they cached themselves, but a poor rate for seeds they saw their neighbor store, even just six minutes before.

Chickadee families of five to nine members often stay together through the winter and join other chickadees, Red-breasted and White-breasted Nuthatches, Downy and Hairy Woodpeckers, Brown Creepers, and Golden-crowned Kinglets to form "foraging flocks." Moving through the woods, these flocks stop to feed at productive sites. Centered around one dominant mated chickadee pair, the flock will defend nearly 20 acres from other roving flocks. Since all other chickadees are submissive to the nuclear pair, they eliminate energy-wasting infighting. All know their place. The advantage of the foraging flock to nonchickadee species may be the Blackcaps' early predator-warning system. Escaping capture by a Northern Shrike, Sharp-shinned Hawk, or Saw-whet Owl becomes easier for birds hiding within the flock.

To survive the long winter nights, which may last eighteen hours at the northern edge of their range, chickadees possess the incredible ability to go into a mild state of torpor. A drop of 20° brings the sleeping chickadee's body temperature to 88° F, lowering its metabolic rate and thereby conserving precious energy. Roosting in tree cavities or in dense evergreens also aids in survival by reducing body heat lost to the air.

Sparky says: Imitate a chickadees's whistle. The high whistled "feeebeee," the Black-capped Chickadee's true song, serves to attract a mate and defend a territory. Imitate the whistle. Can you get a chickadee to answer you?

Fast and furious

Chickadee wings beat thirty strokes per second allowing them to change flight direction in 3/100 of a second. That's a fast U-turn!

Winter Wren
Troglodytes troglodytes

Small brown bird with stubby erect tail

Length 4 inches

Song is a loud, long liquid series of trills and warbles

The Winter Wren is a mighty mite, possessing a huge voice that emanates from a tiny body. Boundary Waters and Quetico paddlers can hear the wren's song from mid-May until the end of July. Once you hear it, the call is hard to forget. One author described the song as a "tumbling series of liquid notes." I timed one individual's songs in May 1991 and found them to last from four to nine seconds, with the average being seven or eight seconds. No other North Woods bird performs a single song as long as the Winter Wren's.

Roger Tory Peterson describes the song as "a rapid succession of high tinkling warbles and trills . . . often ending on a very high light trill." Don't think I'm copping out, but I agree with Roger. For music majors I include the quote that marvels at the Winter Wren's song, which "when played back at one half or one quarter speed reveals a remarkable blend of half-tones and overtones all sung at the same time." The high notes have also been likened to those of a tiny piccolo.

Despite the prominence of the Winter Wren's voice in the boreal landscape, the bird itself is rarely seen. It lives a mouselike existence in the tangled growth of the forest floor.

Averaging 4 inches long, the Winter Wren is the third smallest bird in eastern North America, outsizing only the Ruby-throated Hummingbird (3 1/2 inches) and the Golden-crowned Kinglet (3 3/4 inches).

Truly a little brown bird, the wren is highlighted by a cinnamon back, buffy undersides, a lighter line above the eye, and the stubby, often erect, wren family tail.

The Winter Wren is the only member of the wren family found in Europe, where it is known simply as "the Wren."

My nest or yours?

The male builds a crude "dummy" nest and displays near it in hopes of attracting a mate . . . or two. Researchers believe that Winter Wrens are polygynous, which means the male mates with more than one female. The display to attract females to his "love nest" involves wing quivers, side-to-side tail flicking, then a short call, after which he erects the back feathers and fans his wings.

I think I'm in love

Once a pair bond forms, both sexes build a real nest, often hidden deep in the tangled roots of an upturned tree. The cavity is filled with moss and lined with hair and feathers. Even veteran nest finders like Hal H. Harrison, author of *Peterson's Field Guide to Birds' Nests,* find it nearly impossible to pinpoint the Winter Wren's abode even if narrowed down to one rock crevice, one root mass, or a single mossy hummock.

 **Sparky says:** Before your trip, listen to a tape of bird songs to familiarize yourself with the distinctive songs of the canoe country. Concentrate on learning the distinctive songs of the Common Loon, Broad-winged Hawk, Merlin, Pileated Woodpecker, Least Flycatcher, Common Raven, Black-capped and Boreal Chickadee, Winter Wren, Hermit and Swainson's Thrush, Veery, Ruby-crowned Kinglet, Red-eyed and Solitary Vireo, all the warblers, Dark-eyed Junco, Chipping and White-throated Sparrow. Make notes to yourself in your birdbook. Since many of these birds are difficult to see, knowing their song will greatly aid in identification and your enjoyment.

Time the songs of the Winter Wren. What was the longest call? Were any over eight seconds? Calculate the average duration of one song.

Swainson's Thrush

Catharus ustulatus

Slightly smaller than robin

Gray-olive back, spotted buffy breast

Reedy upward-spiraling song heard mainly at dusk

You finish the supper dishes just as the sun drops behind the horizon's spruce curtain. Suddenly the mosquito hordes emerge, accompanied by the beautiful reedy upward-spiraling song of the Swainson's Thrush. You dive for the tent, zipping out the winged marauders, and lie back to enjoy the evening serenade of English naturalist William Swainson's thrush—canoe country vespers to drain the day's strain.

The Swainson's may be joined by its thrush cousins, the Veery, Hermit Thrush, and American Robin, in the evening chorale. The breezy song of the Veery spirals down the scale. The Hermit Thrush's song consists of several phrases, each given at a different pitch and starting with a long introductory note. Ethereal and serene, say some. I agree. The backyard robin, also a thrush, is equally at home in the spruce-moose forests of the north, where it sings morning, noon, and night. Even though all four thrush songsters breed in the canoe country, it is unlikely that they will all perform at one campsite. Veeries prefer deciduous woods, Swainsons are partial to coniferous forests and bogs, and American Robins and Hermit Thrushes happily live in either habitat.

Formerly christened the olive-backed thrush, the Swainson's does have a gray-olive back, a spotted buffy breast, and light eye ring. The Veery is warm cinnamon above, with a buffy chest and indistinct breast spotting. The Hermit Thrush has characteristics of both, possessing a rusty tail, olive back, and spotted breast. All are robin-shaped but a head smaller than their larger cousin. Shy and retiring, the boreal thrushes are more easily heard than seen, so it makes more sense to learn their songs than to fret over the size of their breast spots.

Sparky says: Study the craftsmanship of a woven bird nest. The Swainson's Thrush builds its nest on a horizontal limb close to the trunk of a smaller tree. Twigs and grass woven together with moss, rootlets, leaves, cedar, birch bark, animal hair, lichens, and skeletonized leaves are the soft, comfy materials that line the well-shaped cup. Now try to build your own. Use litter from the forest floor to create your own personalized nest. Mud, twigs, bark, grass, moss, hair, roots, lichens, and leaves are all legal building supplies. No picking of live material allowed. Who can create the most symmetrical nest cup? Whose nest is the strongest? Whose is the prettiest? Share your nesting attempt with the whole group.

Cedar Waxwing

Bombycilla cedrorum

Length 7 inches

Golden brown with a yellow belly

Brown crest stands above a black face mask

Red "wax" globs tip wing feathers

Voice a high thin wheezy whistle

Like insecure hoodlums, Cedar Waxwings rarely appear in public without a gang of friends about them. Even in summer, when most birds are busy defending territories or taking care of the nest, waxwings find time to gather with three or four friends for a gab session. Their high and thin wheezy voices announce their presence to campers.

In summer, the season of abundant bugs, waxwings modify their vegetarian winter ways to become meat eaters. Perched on exposed branches, Cedar Waxwings sally forth to nab passing mosquitoes and moths, a meager meal but meat nonetheless. They return to the exact perch to await the passing of more flying food. You can easily observe this behavior from a canoe.

The ripening berries of late summer and fall lure the waxwings away from their carnivorous ways and back to a vegetarian existence. Ripe red Mountain Ash berries seem to be a particular favorite. Large flocks congregate on berry-laden trees, stripping them clean in a matter of hours. Cedar Waxwings head south by late fall and are replaced by the more northerly Bohemian Waxwings, which arrive in the Boundary Waters and Quetico in late October.

The Cedar Waxwing is a 7-inch-long golden brown bird with a yellow belly, a sleek black mask, and a real Dippity-Do crest. Its secondary flight feathers appear to have been dipped in red sealing wax, while its tail appears to have been dipped in yellow. Its voice is a high thin lisp.

The Bohemian Waxwing is a slightly longer and portlier version of the Cedar Waxwing, with a deeper voice. And as its name suggests, it wears more gaudy and colorful raiment. Overall golden brown, without the Cedar's yellow belly, the Bohemian makes up for it by sporting bright rusty undertail coverts and lots of "jewelry" in the form of wing decoration. Both species look very well-groomed and finely manicured.

The seal of approval

The red globules that tip the secondary feathers on each wing of the adult Cedar Waxwing are, of course, not wax at all but rather the fused shaft and outer vane of the feather, pigmented red. Red globules on the wing tips denote age and status and are not acquired until the ripe old age of three. D. James Mountjoy and Raleigh Robertson of Queens University scrutinized the sex lives of a flock of waxwings and discovered that those with more red "jewelry" were older, higher-ranking birds that succeeded in life. They mated earlier, laid more eggs, and fledged more offspring than younger birds with little, if any, red jewelry.

A feather in your hat

Women's millinery fashions could have sounded the death knell for the Cedar Waxwing. On a casual stroll down Manhattan streets in 1886, ornithologist Frank Chapman counted the feathers from forty bird species adorning the women's hats displayed in shop windows. The most common feather was the waxwing's, which he found attached to twenty-three different hats. Fortunately, the Lacey Act of 1900 made illegal the interstate transport of birds killed in violation of state laws. Even more inclusive, the historic 1918 Migratory Bird Treaty protected all migratory nongame birds. It would have been tragic to lose the last waxwing for just a few feathers to stick in a hat.

Husband abuse

The Koyukon people of central Alaska talk of the "Distant Time" when the waxwing had a deep voice but no crest. Mr. Waxwing had a jealous and overbearing wife, who dragged him around by the hair, making him cry out until his voice was nothing but a thin squeak. And this was not all—his hair never returned to its former way and still sticks up today. This is how the Cedar Waxwing got his crest and high voice.

Sparky says: Sway in the arms of a breeze-tossed old White Pine. It is an enchanting and slightly scary experience. Sitting high in a tree's branches lets you see the world from the avian perspective. Clamber up an easy-to-climb tree and sit and observe. Few animals look up high for predators. You only need to be 6 to 10 feet off the ground for this camouflage to be effective.

Red-eyed Vireo

Vireo olivaceus

Length 6 inches

Olive-back, blue-gray crown

Red eyes

Song is "vireo . . . varit . . . varee"

You may never see the red eyes of this arboreal vireo, but its nonstop song is sure to attract your attention. Listen for the Red-eyed Vireo in areas with large aspens and birches. The male, perched high in the trees, sings from dawn to dusk. Without binoculars it is nearly impossible to locate the still, sky-high songster. But once you commit to memory its song, aural identification will be a snap. The song consists of a series of ascending and descending two-to-three-syllable phrases interspaced with short pauses. "Vireo . . . varit . . . varee." It is most likely that the name "vireo" arose from a phrase in its song. Red-eyed Vireo has a much nicer ring to it than Red-eyed Varit, don't you think?

On the next sweltering August afternoon, when all the canoe country world seems to be napping, don't be surprised to hear the tireless song of the Red-eyed Vireo wafting down from the crown of an eighty-year-old birch.

The Red-eyed Vireo's most distinctive field mark, its blood-red eye, distinguishes it from fellow canoe country vireos, the Philadelphia and Solitary Vireos. But this feature is hard to detect except at close range. You'll find it easier to recognize this 6-inch-long bird by its olive back, white undersides, gray cap, and white stripe above the eye, bordered in black.

The Solitary Vireo wears spectacles of white, has two white wing bars, and lives in the wetter Black Spruce—Tamarack community. In other words, it haunts the deep, dark, and distant bogs. But it also inhabits upland Jack Pine forests, where it overlaps with the Red-eyed Vireo. Its song is similar to the Red-eyed Vireo's, but bird-book authors describe it as "higher and sweeter."

All vireos resemble warblers but are slightly larger and have stout, short, and slightly hooked bills. The "mini-shrike" bill makes a perfect predator tool for pursuit and capture of larger insects such as spiders, caterpillars, and beetles. Vireos also differ from warblers in their sluggish treetop foraging. Warblers flit hurriedly from twig to twig; vireos move with deliberate motions, like a warbler in slow motion.

Endurance singer

Male birds have an inborn instinct to sing during the breeding season to attract a mate and defend a territory. But this instinct often weakens as the breeding season wears on and the young hatch. Song intensity is strongest in the morning and to a lesser degree in the evening, with a noontime siesta period. But the male Red-eyed Vireo will sing when others won't; he is often the only midday singer. Hot days, cold days, mornings, evenings, high noons, June, or September, it makes no difference to the enthusiastic vireo. In fact, the Red-eyed Vireo holds the continental singing endurance record. Ornithologist Louise Lawrence sat for fourteen hours in an Ontario woodland counting the 22,197 songs given by a single male! That is nearly thirty songs every minute all day long. This incredible song pace has been shown to increase to sixty per minute at the start of the female's incubation period. The Red-eyed Vireo feels the need to sing so strongly that it will even belt out a song when struggling to capture a large insect.

Once believed to be eastern North America's most common breeding bird, today the Red-eyed Vireo is declining because of a triple threat. Forest fragmentation development in the eastern U.S. is the first. A more hidden threat to the Red-eyed Vireo occurs right in its own home—brood parasitism. The ever-increasing Brown-headed Cowbird does not build a nest, but instead lays its eggs in the nests of other species, leaving the often much smaller parent species to raise the cowbird chick at the expense of their own young. This is an adaption from the days when Brown-headed Cowbirds followed the ever-moving herds of bison across the American West to feed on body ticks and on insects kicked up by their passage. Since the birds never stayed in one spot for long and could not take time to build a nest and raise young, they let total strangers take responsibility for raising their offspring. Cowbirds are known to parasitize the nests of over 200 North American species. Host species can be divided into "acceptor" and "rejector" species. Rejectors, such as the American Robin, Blue Jay, Gray Catbird, and Brown Thrasher recognize the alien egg and eject it from the nest. Acceptors accept the egg as one of their own and incubate it till hatching, then feed the cowbird nestling till it fledges. Warblers, Eastern Phoebes, Song Sparrows, and the Red-eyed Vireo are all acceptors.

The Red-eyed Vireo may be the most common recipient of the cowbird's wanton egg laying. Parasitism rates in various studies range from 40 to 75 percent of all vireo nests! William Southern discovered that 75 of 104 Red-eyed Vireo nests in Cheboygan County, Michigan, were parasitized. In such nests it is very fortunate if any of the vireo's own offspring survive to fledge. The cowbird has the advantage, as it hatches sooner, grows faster, and is much larger than its nest mates. Most vireo chicks cannot compete and eventually starve to death.

The major reason for the vireo's decline, though, may be the destruction of their winter home in the tropical forests of the Amazon Basin. Those forests supply the vireos with their support for eight months of the year. As acres fall, fewer and fewer Red-eyed Vireos find a winter home and hence fewer return to the forests of North America. The inevitable result is that we will hear less of the "voluble singer of the tree-tops" in our own canoe country.

 Sparky says: From the comfort of your campsite, count the number of songs given by a Red-eyed Vireo at different times of the day. Count for ten minutes at 6:00 A.M., 9:00 A.M., Noon, 3:00 P.M., 6:00 P.M., and 9:00 P.M. Record the results on a chart. Do any patterns emerge? June and the first half of July are the best times to conduct your song survey.

Warblers enliven summer forests all across North America. Beautifully patterned and colorful, our fifty-one species of breeding warblers wear plumage in yellow, white, black, red, orange, gray, green, gold, chestnut, rufous, blue, brown, and buff—truly the crown jewels of American songbirds. They top the "wanted" list for many birders visiting from overseas.

The North Woods, and specifically the Boundary Waters–Quetico, offers one of the most "warbler-rich" spots on the continent. Twenty-four species breed here. Two species, the Blackpoll Warbler and Orange-crowned Warbler, merely pass through on their way to more northerly breeding grounds. Theoretically, on a one-week canoe trip in early June, an observant (and lucky) canoeist may see over half of North America's warbler species. Unfortunately, many are secretive, buried in brush or high in the tree tops, with only their distinct songs to give them away.

Almost exclusively insectivorous, warblers visit us in the north only for the "bug months." Come late August and September they begin to wing south toward Mexico and Central and South America, where they can still find "flying food." There they spend six to seven months.

This schedule brings up an interesting point. Are these North American warblers that winter in the south, or are they tropical birds that only go north for three months to raise their young and exploit the abundant insect life? The question may be moot if tropical deforestation continues at its current rate. Slashing and burning the rain forest to graze cattle so Americans can have cheap hamburgers and meaty dog food is a travesty. Not only do wintering warblers lose, but a whole ecosystem may be destroyed, leading to inevitable worldwide consequences.

Megamigration

A several-thousand-mile migration is no picnic for this sprite of a bird. Researchers have discovered that migrant warblers have

Warblers
Family *Emberizidae*

Yellow-rumped Warbler

Tiny, colorful, and fast moving

24 species of warblers flit through the North Woods

Feed nearly exclusively on insects

Blackburnian Warbler

Black-and-white Warbler

wings designed for sustained flight, being longer and more pointed than those of year-round tropical warblers that fly only short distances.

The most amazing warbler migration has to be that of the Blackpoll Warbler. They fly nonstop over open ocean from the northeastern United States/southeast Canada to their wintering grounds in northern South America. How do they fuel up for a 4-day nonstop flight? Breeding in America's coniferous forests, Blackpolls gather on the east coast for a ten-to-twenty-day feeding frenzy in which they double their body weight with stored fat. This 1/3 ounce of fat is aviation fuel for 105 to 115 hours of warbler flight at the burn rate of .5 percent of body weight per hour. In spring Blackpolls forsake the open water course, preferring to migrate north overland.

Niche

Carnivores! Warblers are carnivores. They all eat insects and insects are meat, though mosquito steaks are nothing to brag about. How do twenty-four species of warblers, all in search of insects, survive in the canoe country? The answer is niches. Each warbler occupies a unique microhabitat that supplies its food requirements. Tennessee Warblers glean the foliage of mature deciduous trees. Chestnut-sided Warblers prefer second-growth trees. Black-and-white Warblers creep along branches and up and down trunks searching for insects in the bark. Large Red and White Pines are the niche occupied by the appropriately named Pine Warbler. Ovenbirds, like bums searching for a cast-off cigarette butt, methodically walk the forest floor in quest of bugs. Streamsides are where the Northern Waterthrushes feed.

But what about the Cape May, Yellow-rumped, Black-throated Green, Blackburnian, and Bay-breasted Warblers that all feed in spruces? A basic ecology principle holds that when two or more species occupy the same niche, one will lose out. The classic work of Robert MacArthur revealed an amazing story of microniches within a single tree. He discovered that each species preferred to forage in a different part of the tree. The Cape May stayed mainly on the outside top of the tree, moving vertically, sallying forth occasionally to hawk an insect. The Blackburnian also fed on the outside top, but foraged lower down on the tree, moved horizontally, and rarely hawked insects. The spruce's middle interior was the niche of the Bay-breasted, while the Yellow-rumped foraged on the bottom foliage and used the widest variety of food-gathering techniques. MacArthur also found that their nesting periods varied slightly so their peak food demands were slightly offset. Five species that at first glance appeared to compete actually occupy slightly different niches in order to share a limited resource. Fantastic!

 Sparky says: Get familiar with warblers' songs by listening to tapes before a trip. It takes years, though, to master warbler songs. Remember that male warblers sing to attract a mate and defend a territory. When the nesting season is over in late July, so is the singing. Armed with a good pair of binoculars and field guide, quietly stalk the morning woods around your campsite. Do some "pishing" to bring them in for a closer look. Have fun, and don't let frustration ruin the experience.

Budworm boom and bust

Irregular outbreaks of Spruce Budworm are a boon to Bay-breasted, Tennessee, and Cape May Warblers, who feed heavily on the caterpillar. But because of their dependence on such an unstable food supply, these warblers regularly lay five to seven eggs, compared to the four of most warblers. In years of abundance, parents can feed all the young, allowing them to survive to fledge. During low Spruce Budworm years, very few of the warbler chicks may survive.

White-throated Sparrow

Zonotrichia albicollis

Brown body

Black and white striped head

White throat

Song is a whistled "Old man Peabody Peabody Peabody"

Like smoke from a long-out forest fire, the high clear-whistled song of the White-throated Sparrow wafts across the still waters. In the gathering darkness, from the tip of one small Jack Pine in a sea of many, he throws his head back and sings for all the world to hear, or at least for all the other Whitethroats to hear. The burn, from where he and many others sing, dates back to the 1976 drought. It has become hundreds of acres of Quaking Aspen suckers, head-high Jack Pine and scattered snags. Here the sparrow finds food, shelter, and excellent song perches.

The North Woods canoeist sees few but hears many White-throated Sparrows. Singing from nearly any habitat, but particularly from the new growth of recent burns, the bird seems to find morning, noon, evening, and dead of night all acceptable times for song. I have, on more than one occasion, been awakened from deep, middle-of-the-night slumber by an ambitious male trying to defend his territory or woo a female through song. Could they possibly be sleep-singing?

To some campers, every sparrow is just a "little brown bird," worthy only of disdain or disregard. But the sparrows make up a diverse and beautiful group, colored in gold, black, chestnut, buff, olive, and red, and occupying bogs, prairies, tidal marshes, sagebrush flats, tundra, farmlands, mountains, and woodlands.

Adult breeding White-throated Sparrows are striking. Their pure white throat and distinct black-and-white head stripes stand in stark contrast to the brown back and even gray undersides. A yellow spot lies between the eye and the bill. A tan-striped form, with black and gray head stripes, can sometimes be found. Immature and winter adults resemble the tan-striped form. Because of this, and the fact that the sexes look alike, it is impossible to tell age or sex by plumage.

The Whitethroat's song imprints itself on the listener. It has two ascending or descending introductory notes followed by three triple notes all loudly whistled. Many represent the song with the words "Old man Peabody Peabody Peabody."

Pure sweet Canada

More than even the Common Loon's call, the song of the White-throated Sparrow takes me back to the canoe country and brings with it a flood of memories.

Two versions exist of the high clear-whistled song. One starts low and goes high and the other starts high and goes low. Both are used to defend the small home territory and in early summer to attract a female. I have already stated that the Whitethroat sings "Old man Peabody Peabody Peabody," but the sparrows across the border in the Quetico sing "Pure sweet Canada Canada Canada" in honor of their homeland.

En route to Hudson Bay by canoe, some friends once encountered a gospel-singing White-throated Sparrow. Camped on a sandbar on the upper reaches of the Winisk River, they heard an unfamiliar birdsong. Upon investigation they found the familiar Whitethroat whistling the last part of the second line of "When the Saints go Marching in!"

Numerous White-throated Sparrows may be heard singing from a single small burn or logged area.

Sparky says: When you find yourself close to a singing male, try whistling an imitation of his song. Breeding Whitethroat males counter-sing with their neighbors to answer challenges to their territorial rights. If you're good, he may counter-sing with you. If you don't make a good Whitethroat, you'll most likely be ignored as just another strange species in his woods.

Birds of the Boundary Waters and Quetico

*   migrant only
**   winter visitor

Family Gaviidae: Loons
- ❏ Common Loon   *Gavia immer*

Family Podicipedidae: Grebes
- ❏ *Red-necked Grebe   *Podiceps grisegena*
- ❏ *Horned Grebe   *Podiceps auritus*
- ❏ Pied-billed Grebe   *Podilymbus podiceps*

Family Phalacrocoracidae: Cormorants
- ❏ Double-crested Cormorant   *Phalacrocorax auritus*

Family Ardeidae: Herons and Bitterns
- ❏ Great Blue Heron   *Ardea herodias*
- ❏ American Bittern   *Botaurus lentiginosus*

Family Anatidae: Swans, Geese and Ducks
- ❏ *Tundra Swan   *Cygnus columbianus*
- ❏ Canada Goose   *Branta canadensis*
- ❏ *Snow Goose   *Chen caerulescens*
- ❏ Mallard   *Anas platyrhynchos*
- ❏ American Black Duck   *Anas rubripes*
- ❏ *Gadwall   *Anas strepera*
- ❏ *Northern Pintail   *Anas acuta*
- ❏ *Green-winged Teal   *Anas crecca*
- ❏ Blue-winged Teal   *Anas discors*
- ❏ American Wigeon   *Anas americana*
- ❏ *Northern Shoveler   *Anas clypeata*
- ❏ Wood Duck   *Aix sponsa*
- ❏ *Redhead   *Aythya americana*
- ❏ Ring-necked Duck   *Aythya collaris*
- ❏ *Canvasback   *Aythya valisineria*
- ❏ *Greater Scaup   *Aythya marila*
- ❏ *Lesser Scaup   *Aythya affinis*
- ❏ Common Goldeneye   *Bucephala clangula*
- ❏ Bufflehead   *Bucephala albeola*
- ❏ Hooded Merganser   *Lophodytes cucullatus*
- ❏ Common Merganser   *Mergus merganser*
- ❏ *Red-breasted Merganser   *Mergus serrator*

Family Cathartidae: American Vultures
- ❏ Turkey Vulture   *Cathartes aura*

Family Accipitridae: Kites, Hawks, Eagles, and Harriers
- ❏ Northern Goshawk   *Accipiter gentilis*
- ❏ Sharp-shinned Hawk   *Accipiter striatus*
- ❏ Cooper's Hawk   *Accipiter cooperii*
- ❏ Red-tailed Hawk   *Buteo jamaicensis*

❏ Broad-winged Hawk   *Buteo platypterus*
❏ *Rough-legged Hawk   *Buteo lagopus*
❏ *Golden Eagle   *Aquila chrysaetos*
❏ Bald Eagle   *Haliaeetus leucocephalus*
❏ Northern Harrier   *Circus cyaneus*

Family Pandionidae: Osprey
❏ Osprey   *Pandion haliaetus*

Family Falconidae: Falcons
❏ *Peregrine Falcon   *Falco peregrinus*
❏ Merlin   *Falco columbarius*
❏ American Kestrel   *Falco sparverius*

Family Tetraonidae: Grouse and Ptarmigan
❏ Spruce Grouse   *Bonasa umbellus*
❏ Ruffed Grouse   *Dendragapus canadensis*

Family Rallidae: Rails, Gallinules, and Coots
❏ Sora   *Porzana carolina*
❏ *American Coot   *Fulica americana*

Family Gruidae: Cranes
❏ *Sandhill Cranes   *Grus canadensis*

Family Charadriidae: Plovers
❏ *Semipalmated Plover *Charadrius semipalmatus*
❏ Killdeer   *Charadrius vociferus*
❏ *Lesser Golden-Plover *Plurialis squatarola*
❏ *Black-bellied Plover *Plurialis dominica*

Family Scolopacidae: Woodcock, Snipe and Sandpipers
❏ American Woodcock   *Scolopax minor*
❏ Common Snipe   *Gallinago gallinago*
❏ *Whimbrel   *Numenius phaeopus*
❏ Spotted Sandpiper   *Actitis macularia*
❏ Solitary Sandpiper   *Tringa solitaria*
❏ *Greater Yellowlegs   *Tringa melanoleuca*
❏ *Lesser Yellowlegs   *Tringa flavipes*
❏ *Red Knot   *Calidris canutus*
❏ *Pectoral Sandpiper   *Calidris melanotos*
❏ *White-rumped Sandpiper *Calidris fuscicollis*
❏ *Baird's Sandpiper   *Calidris bairdii*
❏ *Least Sandpiper   *Calidris minutilla*
❏ *Dunlin   *Calidris alpina*
❏ *Semipalmated Sandpiper *Calidris pusilla*
❏ *Sanderling   *Calidris alba*

❏ *Ruddy Turnstone   *Arenaria interpres*
❏ *Short-billed Dowitcher   *Limnodromus griseus*
❏ *Long-billed Dowitcher   *Limnodromus scolopaceus*
❏ *Stilt Sandpiper   *Calidris himantopus*
❏ *Buff-breasted Sandpiper   *Tryngites subruficollis*

Family Laridae: Gulls and Terns
❏ *Glaucous Gull   *Larus hyperboreas*
❏ Herring Gull   *Larus argentatus*
❏ *Thayer's Gull   *Larus thayeri*
❏ *Ring-billed Gull   *Larus delawarensis*
❏ *Bonaparte's Gull   *Larus philadelphia*
❏ *Caspian Tern   *Sterna caspia*
❏ *Common Tern   *Sterna hirundo*

Family Cuculidae: Cuckoos and Anis
❏ Black-billed Cuckoo   *Coccyzus erythropthalmus*

Family Strigidae: Typical Owls
❏ Great Horned Owl   *Bubo virgineanis*
❏ **Snowy Owl   *Nyctea scandiaca*
❏ Northern Hawk-Owl   *Surnia ulula*
❏ Barred Owl   *Strix varia*
❏ Great Gray Owl   *Strix nebulosa*
❏ Long-eared Owl   *Asio otus*
❏ Boreal Owl   *Aegolius funereus*
❏ Northern Saw-whet Owl   *Aegolius acadicus*

Family Caprimulgidae: Goatsuckers
❏ Whip-poor-will   *Caprimulgus vociferus*
❏ Common Nighthawk   *Chordeiles minor*

Family Apodidae: Swifts
❏ Chimney Swift   *Chaetura pelagica*

Family Trochilidae: Hummingbirds
❏ Ruby-throated Hummingbird   *Archilochus colubris*

Family Alcedinidae: Kingfishers
❏ Belted Kingfisher   *Ceryle alcyon*

Family Picidae: Woodpeckers
❏ Northern Flicker   *Colaptes auratus*
❏ Pileated Woodpecker   *Dryocopus pileatus*
❏ Yellow-bellied Sapsucker   *Sphyrapicus varius*
❏ Hairy Woodpecker   *Picoides villosus*
❏ Downy Woodpecker   *Picoides pubescens*
❏ Black-backed Woodpecker   *Picoides arcticus*

❏ Three-toed Woodpecker   *Picoides tridactylus*

Family Tyrannidae: Tyrant Flycatchers
❏ Eastern Kingbird   *Tyrannus tyrannus*
❏ Great Crested Flycatcher   *Myiarchus crinitus*
❏ Eastern Phoebe   *Sayornis phoebe*
❏ Yellow-bellied Flycatcher   *Empidonax flaviventris*
❏ Alder Flycatcher   *Empidonax alnorum*
❏ Least Flycatcher   *Empidonax minimus*
❏ Eastern Wood–Pewee   *Contopus virens*
❏ Olive-sided Flycatcher   *Contopus borealis*

Family Alaudidae: Larks
❏ *Horned Lark   *Eremophila alpestris*

Family Hirundinidae: Swallows
❏ Tree Swallow   *Tachycineta bicolor*
❏ Bank Swallow   *Riparia riparia*
❏ Northern Rough-winged Swallow   *Stelgidopteryx serripennis*
❏ *Barn Swallow   *Hirundo rustica*
❏ Cliff Swallow   *Hirundo pyrrhonota*

Family Corvidae: Jays, Magpies, and Crows
❏ Gray Jay   *Perisoreus canadensis*
❏ Blue Jay   *Cyanocitta cristata*
❏ Common Raven   *Corvus corax*
❏ American Crow   *Corvus brachyrhynchos*

Family Paridae: Titmice, Verdins, and Bushtits
❏ Black-capped Chickadee   *Parus atricapillus*
❏ Boreal Chickadee   *Parus hudsonicus*

Family Sittidae: Nuthatches
❏ White-breasted Nuthatch   *Sitta carolinensis*
❏ Red-breasted Nuthatch   *Sitta canadensis*

Family Certhiidae: Creepers
❏ Brown Creeper   *Certhia americana*

Family Troglodytidae: Wrens
❏ House Wren   *Troglodytes aedon*
❏ Winter Wren   *Troglodytes troglodytes*
❏ Sedge Wren   *Cistothorus platensis*

Family Mimidae: Mockingbirds and Thrashers
❏ Gray Catbird   *Dumetella carolinensis*
❏ Brown Thrasher   *Toxostoma rufum*

Family Muscicapidae: Thrushes, Kinglets and Bluebirds
- ❑ American Robin — *Turdus migratorius*
- ❑ Wood Thrush — *Hylocichla mustelina*
- ❑ Hermit Thrush — *Catharus guttatus*
- ❑ Swainson's Thrush — *Catharus ustulatus*
- ❑ *Gray-cheeked Thrush — *Catharus minimus*
- ❑ Veery — *Catharus fuscescens*
- ❑ Eastern Bluebird — *Sialia sialis*
- ❑ Golden-crowned Kinglet — *Regulus satrapa*
- ❑ Ruby-crowned Kinglet — *Regulus calendula*

Family Motacillidae: Pipits
- ❑ *American Pipit — *Anthus spinoletta*

Family Bombycillidae: Waxwings
- ❑ **Bohemian Waxwing — *Bombycilla garrulus*
- ❑ Cedar Waxwing — *Bombycilla cedrorum*

Family Laniidae: Shrikes
- ❑ **Northern Shrike — *Lanius excubitor*

Family Sturnidae: Starlings
- ❑ European Starling — *Sturnus vulgaris*

Family Vireonidae: Vireos
- ❑ Solitary Vireo — *Vireo solitarius*
- ❑ Red-eyed Vireo — *Vireo olivaceus*
- ❑ Philadelphia Vireo — *Vireo philadelphicus*
- ❑ Warbling Vireo — *Vireo gilvus*

Family Emberizidae: Wood Warblers, Sparrows, Blackbirds and Finches
- ❑ Black-and-white Warbler — *Mniotilta varia*
- ❑ Golden-winged Warbler — *Vermivora chrysoptera*
- ❑ Tennessee Warbler — *Vermivora peregrina*
- ❑ *Orange-crowned Warbler — *Vermivora celata*
- ❑ Nashville Warbler — *Vermivora ruficapilla*
- ❑ Northern Parula — *Parula americana*
- ❑ Yellow Warbler — *Dendroica petechia*
- ❑ Magnolia Warbler — *Dendroica magnolia*
- ❑ Cape May Warbler — *Dendroica tigrina*
- ❑ Black-throated Blue Warbler — *Dendroica caerulescens*
- ❑ Yellow-rumped Warbler — *Dendroica coronata*
- ❑ Black-throated Green Warbler — *Dendroica virens*
- ❑ Blackburnian Warbler — *Dendroica fusca*
- ❑ Chestnut-sided Warbler — *Dendroica pensylvanica*
- ❑ Bay-breasted Warbler — *Dendroica castanea*

❏	*Blackpoll Warbler	*Dendroica striata*
❏	Pine Warbler	*Dendroica pinus*
❏	Palm Warbler	*Dendroica palmarum*
❏	Ovenbird	*Seiurus aurocapillus*
❏	Northern Waterthrush	*Seiurus noveboracensis*
❏	Connecticut Warbler	*Oporornis agilis*
❏	Mourning Warbler	*Oporornis philadelphia*
❏	Common Yellowthroat	*Geothlypis trichas*
❏	Wilson's Warbler	*Wilsonia pusilla*
❏	Canada Warbler	*Wilsonia canadensis*
❏	American Redstart	*Setophaga ruticilla*
❏	Red-winged Blackbird	*Agelaius phoeniceus*
❏	Northern Oriole	*Icterus galbula*
❏	Rusty Blackbird	*Euphagus carolinus*
❏	*Brewer's Blackbird	*Euphagus cyanocephalus*
❏	Common Grackle	*Quiscalus quiscula*
❏	Brown-headed Cowbird	*Molothrus ater*
❏	Scarlet Tanager	*Piranga Olivacea*
❏	Rose-breasted Grosbeak	*Pheucticus ludovicianus*
❏	Indigo Bunting	*Passerina cyanea*
❏	Evening Grosbeak	*Coccothraustes vespertinus*
❏	Purple Finch	*Carpodacus purpureus*
❏	**Pine Grosbeak	*Pinicola enucleator*
❏	**Hoary Redpoll	*Carduelis hornemanni*
❏	**Common Redpoll	*Carduelis flammea*
❏	Pine Siskin	*Carduelis pinus*
❏	American Goldfinch	*Carduelis tristis*
❏	**Red Crossbill	*Loxia curvirostra*
❏	**White-winged Crossbill	*Loxia leucoptera*
❏	Savannah Sparrow	*Passerculus sandwichensis*
❏	Dark-eyed Junco	*Junco Hyemalis*
❏	*American Tree Sparrow	*Spizella arborea*
❏	Chipping Sparrow	*Spizella passerina*
❏	Clay-colored Sparrow	*Spizella pallida*
❏	*Harris' Sparrow	*Zonotrichia querula*
❏	*White-crowned Sparrow	*Zonotrichia leucophrys*
❏	White-throated Sparrow	*Zonotrichia albicollis*
❏	*Fox Sparrow	*Passerella iliaca*
❏	Lincoln's Sparrow	*Melospiza lincolnii*
❏	Swamp Sparrow	*Melospiza georgiana*
❏	Song Sparrow	*Melospiza melodia*
❏	*Lapland Longs	*Calcarins lapponicus*
❏	*Snow Bunting	*Plectrophenax nivalis*

Fish

Walleye

Stizostedion vitreum

Average 2 to 8 pounds

Color ranges from dark silver to yellowish gold to olive brown

Large spiny dorsal fin

White patch at bottom of tail fin

Luminescent white eyes

The Walleye, occasionally referred to as the "walleyed pike," is not a pike at all, but rather America's largest perch. Adults average from 2 to 8 pounds. But fish continue to grow throughout their lives, so lunkers are lurking. The Minnesota state angling record is a 17 pound, 8 ounce beauty hooked on opening weekend 1979 in the Seagull River bordering the BWCAW. Wow! LeRoy Chiovitte gives the details of his trophy catch in Joe Fellegy's *Classic Minnesota Fishing Stories.*

Walleye coloration varies from dark silver to yellowish gold to dark olive brown. The fish has a large and distinctive unspotted spiny dorsal fin that makes it look like a miniature sailfish. It also displays a white patch at the bottom of its tail fin. The very similar Sauger lacks these fin characteristics. What really sets the Walleye apart, however, is its white, translucent, lucid, almost luminescent eyes. If you've ever seen one live and in person you know what I'm talking about. They border on eerie.

The 1965 Minnesota state legislature deserves commendation for its choice of the Walleye as the state fish. Native to many clean, high-nutrient lakes in northern and central Minnesota, the Walleye is a major tourist attraction as well as symbol of the Land of Sky Blue Waters.

The Walleye is not native to many parts of the Boundary Waters and Quetico. In the 1920s extensive and random dumping of Walleye fry into any and every lake, pond, and puddle established them in many border country lakes. But in many more lakes the conditions were much too oligotrophic (nutrient-poor), and the Walleye did not survive. The boom business of fly-in fishing spurred the stocking and ultimately may have been responsible for the overfishing of some lakes.

Cannibal cousins

Ice-out in the border country, which can come anytime from mid-April to mid-May, is soon followed by the Walleye spawn as waters reach the 42° F to 52° F range. "Homing" plays an important role in their return to traditional spawning sites. Some make mass migrations up streams and rivers to smaller lakes and tributaries, while others lay eggs on shallow bars or shoals of their home lakes.

The males arrive several days before the females, who, upon arrival, may be attended by more than one male Walleye. He fertilizes (or they fertilize) the eggs as the female lays them, which may be a formidable job—a 5-pound female can lay about one hundred thousand eggs. The adults stay on the spawning grounds for up to a month; then they leave their eggs and resulting young to fend for themselves.

By the time the Walleyes have grown to the fingerling stage (finger length), the lake's perch are just hatching, and they become the meat-and-potatoes diet of the growing "walleyettes." I've heard of kissing cousins, but never cannibal cousins like these two species appear to be. Even as adults, Walleye find smaller perch a real taste treat.

Translucent and luminescent: the Walleye's eye

The word "walleye" is actually derived from an old Norse word meaning "a light beam in the eye." The eye seems to glow because of a special light-sensitive layer of cells called the tapetum lucidum, which is located behind the color-sensitive cone cells and reflects light back out to the cones. This allows the fish a second chance to catch available light and enables Walleyes to see, and hence feed, in the dim light of turbid lakes and during the dawn and dusk hours.

Recently, scientists discovered that the Walleye eye harbors the largest color-sensitive cone cells of any animal in the world. It would take twenty-five human cone cells laid side by side to equal the width of a human hair, while only five Walleye cone cells would be required to span the same distance. The cells are just large enough to implant with the tiniest of microelectrodes. And that is exactly what Dr. D. A. Burkhardt of the

Wandering Walleyes

Radio-tagged Walleyes have been found to travel over 100 miles in a single month.

University of Minnesota is doing in his study of human cone cells in relation to color and fine pattern vision. A flash of light is transmitted to the Walleye, which generates an electrical current in the cone cells that can be measured by the implanted microelectrode. Through this process, Dr. Burkhardt and his associates have discovered that Walleyes have many large orange-sensitive cone cells, far outnumbering the smaller green-sensitive cones. Humans have three cone types that are sensitive to blue, green, and yellow. Walleye can distinguish red from green, but not blue from green. Further research will study how human eyes can adapt nearly instantaneously from darkness to bright light. And with the Walleye eye available to study, new discoveries about the human eye should be right around the corner.

Sparky says: Share the memories of your canoe trip in a letter to a special youngster. Did you catch a Walleye? Did you portage farther than ever before? What is the area like now? What are your hopes for its future? Relate the day-to-day events of the journey. Hold on to the letter. Mail it when the young person turns eighteen.

Jackfish, pickerel, northern, snake pickerel, hammer-handle, and pike all refer to the one and only *Esox lucius,* the great Northern Pike. Found in much of eastern North America's waters, from warm and weedy to clear, deep, and cold, the Northern Pike is also one of only two species of freshwater fish to inhabit the world's three great circumpolar continents: North America, Europe, and Asia.

Built somewhat like a two-by-four, the Northern has a long powerful body that tapers to a duck-bill-shaped head chock full of canine ripping and tearing teeth. The dorsal fin is far back on its body. Its sides and back are bluish green, dappled with light oval spots. The belly is creamy white. Only its larger cousin, the mighty Muskellunge, could be mistaken for the Northern Pike.

Rather than wearing polka dots like a Northern, Muskies sport dark tiger-stripes on a silvery background, which is much more appropriate for a fish that may reach 100 pounds (if you believe the banter of some Lake of the Woods fishing guides). Northern Pike can grow to be monsters too, as evidenced by the 1929 angling record of a 45 pound, 12 ounce behemoth taken on Basswood Lake, in waters shared by the Boundary Waters and Quetico.

Spawn time

Like the Walleye, the Northern spawns soon after the ice goes out. Some ascend small streams to lay eggs; others prefer the grassy margins of lakes. A female, as long as your arm and weighing 5 pounds, lays about sixty thousand eggs and then abandons them. Fertilized eggs hatch in about two weeks, and the tiny Northerns start on baby food: tiny crustaceans and other microscopic critters (zooplankton).

In the fingerling stage they may switch to sucker fry, which hatch at about the same time as the Northerns but are smaller. If food is scarce they become bite-size cannibals, eating their own siblings. That is precisely why Northerns are not raised in

Northern Pike

Esox lucius

Long slender fish averages 5 to 10 pounds

Duck-bill shaped head with sharp teeth

Greenish body dappled with light spots, white belly

fish hatcheries. It is easier to protect their natural spawning areas than to keep them from eating each other in the confines of hatcheries.

Freshwater tyrant

As Izaak Walton, the great fishing philosopher, said back in the mid-1600s, "The mighty luce or pike is taken to be the tyrant of fresh waters." And indeed it is a lean, mean, eating machine. A body built for quick bursts of speed and a mouthful of teeth that even Jaws would be proud of make the lake mates of the Northern a very nervous group indeed. Like a mugger in a big city alley, the Northern hidden by a submerged log or a weed bed lies in wait for its victim. Pity the unsuspecting perch, sucker, sunfish, minnow, smaller Northern, duckling, loonlet, frog, young Muskrat, or leech that passes by and becomes its afternoon snack. When abundant in midsummer, leeches may be breakfast, lunch, and dinner for hungry Northern Pike.

Bluegills must have been abundant in the home waters of a Northern that was speared through the ice and found to have twenty of these fish in its stomach. When the fishing is good, why stop? Northerns don't know about game limits, but they function as the best fisheries managers around. Every ecosystem needs its predators. Northerns control populations of panfish, maintaining them at a healthy level. Without predators, panfish would overrun lakes, eating up food populations and eventually leaving their own numbers stunted.

The myth that these fish don't bite in August because they shed their teeth and have sore mouths is just that, a myth. Northerns do lose teeth, but only over time as they become worn out or broken off, and not all at once. The fish probably don't bite because their waters swarm with so much food in late summer.

Y-bones

Northerns make excellent eating, but many people complain about all those bones. Those little Y-bones are actually extra ribs that are found in many fish but are well developed in pike.

Sparky says: Whatever bait you use, remember to keep it moving. A Northern Pike responds to anything that excites its curiosity. Trolling the shallows and weed beds with live minnows, spoons, plugs, or large bucktail spinners is the tried-and-true method, though I once caught a nice Northern on a vintage lure that was just a stamped piece of metal in the shape of a stereotypical fish. Don't forget those nasty teeth— always use a steel leader.

Lake Trout

Salvelinus namaycush

Average weight 3 pounds

Light spots speckle the gray body

Old, heavy fish develop large bellies

"I bet you're a Pisces"

Adult Lake Trout are piscivores. Omnivores eat "omni," or everything; herbivores eat herbs; and piscivores eat "pisces," or fish. They'll eat nearly any small fish that strays into their dark, deepwater haunts, but will supplement their diet with tasty aquatic insects.

The Lake Trout has been called many things by many people. Early resort owners advertised "landlocked salmon" fishing to lure customers to the North Woods. Commercial fisherman recognized thirty-six breeds of Lake Trout in Lake Superior, including "racers," "blacks," "redfins," "yellowfins," and "bluefins," the latter of which has now been popularized by the Bluefin Bay Condominiums on Minnesota's North Shore. But your typical Lake Trout is slate gray speckled with light spots, and is the shape of a stereotypical fish—you know, like the crayon fish you drew in first grade.

Like couch-potato humans, old heavy trout develop a belly. The largest Lake Trout caught in Lake Superior was a 63-pound fatso brought up from the depths over forty years ago. The Minnesota record belongs to Gustav Nelson of Duluth, who landed a 43 1/2 pound Laker off Hovland in 1955. A photo and his fish story can be found in *Classic Minnesota Fish Stories* by Joe Fellegy. Inland Lakers may top 20 pounds but average about 3.

Lake Trout are native to the deep, cold, boulder-rimmed lakes of the Boundary Waters and Quetico. It needs deep water to escape from the warmer upper waters in summer. Trout thrive only in lake temperatures below 65° F. Rocky shores and shoals provide the ideal birthing center during the late fall spawn. But before they sign a lease on a lake, trout must be sure that dissolved oxygen is plentiful. This usually comes with the territory, as deep, rocky, and cold lakes are botanical deserts, with little plant life to consume available oxygen.

Unlike Walleyes and Northerns, inland Lake Trout spawn in the fall just before the ice forms. On rocky reefs and boulder rubble they randomly scatter their large eggs. The mature female lays only about 6000 eggs (750 per pound of fish), which is about 10 percent of the typical Walleye or Northern spawn. Eggs hatch in sixteen to twenty weeks while beneath the ice, and hatchlings sink or swim alone, so to speak, as the parents play no part in parenting. The young grow slowly in the cold water on a diet of nearly invisible plants and animals.

Commercial fishing

Historically, the Ojibwa, and possibly the Dakota before them, netted Lake Trout for food and profit. The Ojibwa women, using handwoven nettle-stalk twine nets, seined the shallows in autumn for spawning trout, which they dried and smoked for winter use. A medicine charm of Calamus and Wild Sarsaparilla root, dried and powdered, was placed on the nets to attract fish. After each use the women thoroughly cleaned the nets and even dipped them in a sumac-leaf concoction to destroy any remaining fish odor.

With the decimation of North America's Beavers and a switch in European fashions from Beaver-felt hats to silk toppers, American fur baron John Jacob Astor and his American Fur Company dove headfirst into the commercial fishing business on Lake Superior. In 1836 they opened a fishery at the abandoned Grand Portage post of the British-based North West Company, located on the North Shore just south of the present-day international border. Trout were salted and packed in barrels for shipment to Chicago, Detroit, and New York. Profits soon exceeded those of the company's fur trade. But the venture went bust several years later, due in part to the Panic of 1837 and the failure of expansion to distant markets.

The Ojibwa ceded much of their remaining land in 1854, which opened the Lake Superior North Shore to permanent white settlement and nearly closed the door on the traditional Ojibwa way of life. The Norwegian fishermen came first. A rugged, cliff-studded, river-fed, unpredictable, rough-and-tumble, expansive body of spruce- and pine-rimmed northern water reminded them of the old country. The promise of prosperity and a brighter future brought more homesteaders. And the fishing was good. So good, in fact, that word spread, and the number of commercial fishermen on Lake Superior's North Shore increased from fifty in 1890 to more than four hundred in the 1930s. Using rowboats and, later, larger skiffs, fishermen set large mesh gill nets at depths down to 100

Orange meat?

Nothing beats trout fillets or steaks fried golden brown. But have you ever noticed the variation of flesh color from trout to trout, which can range from white to pink to orange to red? Genetics may play a role, but so does diet. A diet primarily of crustaceans can turn trout flesh a pale orange, while a fish feeding heavily on algae-feeding critters may have pinkish red meat.

fathoms and ran them only once or twice a week, the cold water acting to preserve any fish caught. In 1942 three million pounds of Lake Trout were processed and shipped from the U.S. waters of Lake Superior.

But it could not, and did not, last; the collapse came in the 1950s. The two-pronged spear of overfishing and the invading Atlantic Ocean Sea Lamprey struck the fatal blow. Lamprey latch onto trout with a suckerlike mouth, cutting through the skin with razor-sharp teeth to suck out blood and other body fluids. They attack only adults, inflicting a 90 percent death rate. Today, though most of the commercial fishermen have disappeared into the lake fog of history, the Lake Trout is making a comeback as a favorite of sport anglers, charter captains, and tourist boards throughout the Northland.

Sparky says: Know the Lake Trout's temperature requirements. Inland Lake Trout fishing is great sport, summer or winter. But you must follow the trout as they migrate up and down the lake's thermocline through the seasons. Fish that are at 30 feet in June are probably at 70 feet in mid-July, enjoying the cooler waters. Trolling with a spoon and heavy sinker at these depths, you may latch onto the "landlocked salmon," who hits hard and fights well, but seldom leaps.

The early habitants of Quebec called him *achigan,* "the ferocious one," a term borrowed from their Algonquin neighbors, that most likely referred to the Smallmouth's vigorous defense of its nest and brood. But it is also an appropriate name for a fish that many anglers revere as "ounce for ounce the fightingest fish that swims." Aggressive, protective, and a real prizefighter, the Smallmouth is one bad sunfish. Sunfish? Well, sort of. The brown, gold, or olive, but usually dark green, bass with the dark mottling or side barring is a member of the sunfish family (family Centrarchidae) and even has the red eyes to prove it.

Pulling a 4-pounder from the cold, low-nutrient border country lakes makes many anglers hoop and holler and even dance around the campfire, for it is a fine fish. The Minnesota record is an 8-pounder pulled in 1948 from West Battle Lake in Otter Tail County. Eleven-pound "hawgs" have been landed south of the Mason-Dixon line.

Not native to the great North Woods, Smallies were introduced around the turn of the century by the newly arrived lumberjacks who felt that the border country was not complete without them. And the Smallmouth remains long after the last crosscut saw ripped through a towering White Pine. Though native to medium-sized, clear-watered, gravel-bottomed lakes and the Mississippi River drainage of clear, moderately cold, swift streams, Smallmouths have thrived and spread throughout the canoe country lakes to take their place as a top game fish. All attempts to stock the "coffee-colored" streams of the north have failed, though.

Largemouth Bass cannot survive in the cold oligotrophic lakes of the Boundary Waters and Quetico.

The nest-building fish

We do not usually associate nest building with finned fauna, but rather with our feathered friends, the females of which fend off foes. But the Smallmouth Bass crushes stereotypes, a

Smallmouth Bass
Micropterus dolomieui

Average weight is less than 4 pounds

Usually dark green but can be brown, gold, or olive

Mottled or barred sides

Red eyes

Mass bass exodus

Like a bear readying for hibernation, Smallmouth Bass fatten up in the fall in preparation for winter's cold. When the water reaches a cool 50° F, a mass bass exodus takes them to the dark holes and rock crevices of the deeps, where they lie dormant for the long winter. Any person claiming to go ice fishing for bass either is a liar or a real greenhorn, or knows something you don't!

real fish for the modern era and beyond, as it is the father who builds the nest, guards the eggs, and stays home with the kids.

Warming waters trigger the spring spawn. Males arrive first and get the place looking "homey" by building a nest, which is little more than a saucer-shaped depression fanned out by his tail in knee-deep to over-the-head water. Soon a female comes around, and the male courts her by repeatedly chasing her to the nest. How romantic! The female lays her eggs in the nest and the milt of the male fertilizes them. The female is then free to swim away from any parental responsibility and may mate with other males at their nests.

I learned firsthand, or firstleg in this case, how ornery father Smallies can get when they have to stay home with the kids. Swimming near a dock on Seagull Lake several years ago, I was suddenly and repeatedly bumped by a bass that had evidently nested along our private beach. It was a nervy move for a fish who was outweighed 100 to 1, but it shows how aggressive male Smallies can be when it comes to protecting their eggs and hence their own genes.

The eggs soon hatch into wrigglers, which are little more than embryos with an attached yolk sack. They remain in the nest while the father plays watchdog, attacking intruding fish, floating sticks, and even those pesky fishing lures, making him very susceptible to angling at this time of year. As the young mature into free-swimming black fry, they leave the nest behind. But not their parent, who continues to watch over them with a protective eye for another two to three weeks. Finally, as the young develop scales, acquire adult coloration, and can better fend for themselves, they leave the father behind and strike out on their own into the wilds of the great blue under.

Sparky says: Give underwater bass fishing a try on your next canoe trip. All you need is a snorkel and mask, a stick with line, a barbless hook, and a fat worm. Now snorkel down till you find a school of bass, probably near a boulder field, and dangle your baited line in front of them. This was grand sport for us "go-fers" at Tuscarora Outfitters, as we would compete to see who could land the largest Smallmouth in this manner. It was all catch-and-release, of course. Not only is this the ultimate way to fish, because you're in their element, but it is also a fascinating way to learn about bass behavior.

Crayfish cuisine

Everyone knows that Largemouth Bass like frogs . . . to eat, that is. But cousin Smallmouth does not share this culinary affinity, preferring the delicate taste of the orange crustaceans called crayfish. During some parts of the summer and fall, Smallmouth may feed exclusively on them. This explains Smallies' preference for rock-strewn shallows, which are also prime crayfish habitat.

Fish of the Boundary Waters and Quetico

Family Petromyzontidae: Lamprey
- ❏ Silver Lamprey — *Ichthyomyzon unicuspis*

Family Acipenseridae: Sturgeon
- ❏ Lake Sturgeon — *Acipenser Fuluescens*

Family Salmonidae: Trout, Salmon, Char, and Whitefish
- ❏ Lake Whitefish — *Coregonus clupeaformis*
- ❏ Shortjaw Cisco — *Coregonus zenithicus*
- ❏ Cisco/Lake Herring — *Coregonus artedi*
- ❏ Round Whitefish — *Prosipium cylindraceum*
- ❏ Lake Trout — *Salvelinus namaysuch*
- ❏ Brook Trout — *Salvelinus fontinalis*

Family Osmeridae: Smelt
- ❏ Rainbow Smelt — *Osmerus mordax*

Family Umburidae: Mudminnows
- ❏ Central Mudminnow — *Umbra limi*

Family Esocidae: Pike
- ❏ Northern Pike — *Esox lucius*

Family Lyprinidae: Minnows
- ❏ Golden Shiner — *Notemigonus crysoleucas*
- ❏ Finescale Dace — *Phoxinus neogaeus*
- ❏ Northern Redbelly Dace — *Phoxinus eos*
- ❏ Creek Chub — *Semotilus atromaculatus*
- ❏ Pearl Dace — *Margariscus margarita*
- ❏ Lake Chub — *Couesius plumbeus*
- ❏ Blacknose Dace — *Rhinichthys atratulus*
- ❏ Longnose Dace — *Rhinichthys cataractae*
- ❏ Brassy Minnow — *Hybognathus hankinsoni*
- ❏ Common Shiner — *Luxilus cornutus*
- ❏ Fathead Minnow — *Pimephales promelas*
- ❏ Bluntnose Minnow — *Pimephales notatus*
- ❏ Emerald Shiner — *Notropis atherinoides*
- ❏ Mimic Shiner — *Notropis volucellus*
- ❏ Blacknose Shiner — *Notropis heterolepis*
- ❏ Spottail Shiner — *Notropis hudsonius*

Family catostomidae: Suckers
- ❏ White Sucker — *Catostomus commersoni*
- ❏ Longnose Sucker — *Catostomus catostomus*
- ❏ Shorthead Redhorse — *Moxostoma macrolepidotum*

Family Ictaluridae: Bullhead Catfishes
- ❏ Channel Catfish — *Ictalurus punctatus*
- ❏ Tadpole Madtom — *Notorus gyrinus*

Family Percopsidae: Trout-Perches
- ❏ Trout-Perch — *Percopsis omiscomaycus*

Family Gadidae: Cod
- ❏ Burbot — *Lota lota*

Family Gasterosteidae: Sticklebacks
- ❏ Ninespine Stickleback — *Pungitius pungitius*
- ❏ Brook Stickleback — *Culaea inconstans*

Family Cottidae: Sculpins
- ❏ Deepwater Sculpin — *Myoxocephalus thompsoni*
- ❏ Spoonhead Sculpin — *Cottus ricei*
- ❏ Slimy Sculpin — *Cottus cognatus*
- ❏ Mottled Sculpin — *Cottus bairdi*

Family Centrarchidae: Sunfish and Bass
- ❏ Rock Bass — *Ambloplites rupestris*
- ❏ Smallmouth Bass — *Micropterus dolomieui*

Family Percidae: Darters, Perch, Walleyes, and Saugers
- ❏ Walleye — *Stizostedion vitreum*
- ❏ Sauger — *Stizostedion canadense*
- ❏ Yellow Perch — *Perca flavescens*
- ❏ Logperch — *Percina caprodes*
- ❏ Johnny Darter — *Etheostoma nigrum*
- ❏ Iowa Darter — *Etheostoma exile*

Reptiles and Amphibians

Eastern Garter Snake

Thamnophis sirtalis

Average length is 18 to 24 inches

Dark body with three yellow, orange, or red lengthwise stripes

No poisonous snakes inhabit the canoe country. The Timber Rattlesnake is found in southeast Minnesota and southeast Ontario; its cousin rattler, the Massasauga, lives in swampy lowlands of the Minnesota portion of the Mississippi River Blufflands. Rattlers have never inhabited the Boundary Waters and Quetico. The Eastern Garter Snake, though it may become quite large, is entirely harmless except to the toads, frogs, insects, worms, and mice it stalks. Humans most often encounter this snake during its daytime forays along paths and portages. However, on more than one occasion I have seen them swimming.

The world record Eastern Garter Snake measured 48 3/4 inches long, but the average adult size is 18 to 24 inches. Three yellow, orange, or red stripes run the length of the dark snake, one on the back and one down each side. This pattern's resemblance to the old striped garters used to hold up men's socks inspired the snake's name.

The only other Boundary Waters and Quetico snakes are the small Northern Ringneck Snake and the reclusive Red-bellied Snake, which rarely grows larger than 12 inches and has, surprise, a red belly. Humans rarely see them.

Garter snakes, if harassed, will sometimes bite, but often emit a foul-smelling substance from their anal glands as a deterrent to predators.

The snake pits of Manitoba

You'll find the largest concentration of snakes in the world not in the Brazilian jungles, but closer to home, in the interlake region of southern Manitoba. Near the town of Narcisse, north of Winnipeg, limestone sinkholes serve as the winter sleeping quarters for hundreds of thousands of garter snakes. One sinkhole can accommodate ten to fifteen thousand sleepy snakes!

Dr. Michael Aleksiuk found that snakes may travel up to 11 miles to their traditional hibernaculum in Manitoba. This movement usually takes place in the rainy weather of late

September and early October. Joined by up to fifteen thousand of their neighbors, the garter snakes survive the cold Manitoba winter knotted up in a sleeping mass deep in the limestone fissures. Outside temperatures may plummet to -40° F, but the hibernaculum stays a cozy +30° F.

Seven months later, the males venture out into the spring air, one week ahead of the females. When the groggy females emerge from the hibernaculum, each is mobbed by up to one hundred sexually excited males. The ensuing tangled frenzy, known to herpetologists as the "mating ball," is triggered by males picking up the scent of the female pheromone vitellogenin with the scent organs on their rapidly flicking tongues. When one male successfully breeds, he leaves behind a gelatinous plug that gives off a pheromone, repelling other males.

Males wait around the den for another chance to mate. Courtship, if you can call it that, is a fair-weather event. Garter snakes need a body temperature of 77° F to become sexually active. Depending on the number of suitable warm sunny days, the courting season may last from three days to five weeks.

After mating, the mother-to-be splits the scene. Basking in the sun incubates the developing eggs inside her. A dozen to seventy young are born live late in the summer. If they escape Broad-winged Hawks, Mink, and Great Blue Herons, the young garter snakes may survive to carry on the ancient cycle.

Nest lining

Great Crested Flycatchers use the shed skin of the garter snake to line their nest cavities.

Organ donors

Garter snake livers are a true delicacy to the American Crow. One researcher discovered dozens of dead snakes near a hibernaculum, each with two neat holes picked in its body where the livers should have been. They must have been organ donors.

Reptile range

The garter snake is the most widely distributed reptile in North America, ranging from Florida's Everglades to Western Canada.

Sparky says: Match your compass skills against the garter snake's natural ability. Year after year garter snakes migrate up to 11 miles to arrive at their hibernating dens without the aid of compass or map. How do they do it? Lost humans in a wooded environment on a cloudy day often end up traveling in circles. We have no built-in navigational aids. Learn to use a compass and then with its aid bushwhack to a nearby lake. Set a straight line bearing and follow it. Remember, trust your compass and not your senses!

Snapping Turtle

Chelydra serpentina

Large mud-colored turtle

Weight 10 to 35 pounds

Average shell length 8 to 12 inches

Tail nearly length of shell

Warty head, neck, and tail

Nasty-looking hooked jaw

Does not bask in sun

Part-time vegetarians

Contrary to what we may think, Snapping Turtles eat a great amount of vegetation. Their adult diet is about 50 percent aquatic plants and 50 percent animal matter, including insects, crayfish, fish, reptiles, and water birds. They eat ducklings and loonlets, but do not substantially affect populations.

I went skinny-dipping today. I think a lot about Snapping Turtles when I'm skinny-dipping. But my fears are unfounded. Snappers are docile in their home environment, the water. Even a human accidently stepping on them underwater incites only the escape response. When on land, though, watch out. Snappers cannot pull their head and legs into the safety of their shell, like most turtles, and so become very aggressive. They can't bite broomstick handles in half, but they can take a big bite out of your hand.

Turtles never stop growing throughout their long lives. One Snapper lived fifty-seven years in captivity, and some undoubtedly live to be seventy-five years old. They have plenty of time to reach grand size. Adults range from 10 to 35 pounds, with a shell length of 8 to 12 inches. Mud-colored, the Snapper has a large warty wrinkled head that includes a formidable hooked jaw. The claws are long, and the tail is nearly the length of the shell, with horny projections on top. Leeches and algae adorn the elders.

Since Snapping Turtles don't bask on rocks and logs like Painted Turtles, people don't often see them. The canoe country paddler's best chance to see one is in June, when females come on land, traveling quite a distance in some cases, to lay their eggs in sandy banks.

Ping-Pong balls

In June, female Snappers leave the water in search of suitable egg-laying sites. They may swim and walk considerable distances to reach traditional sites. Upon reaching the sandy shore, soft dirt, or loose bank, she excavates a 6-by-6-inch cavity, backs in, and deposits ten to ninety-six round white eggs, nearly the size of Ping-Pong balls. She then covers up the hole, pats down the dirt with her body, and turns away, leaving her offspring to fend for themselves. Most don't survive.

Many end up the way I found a nest one August, raided by a foraging Mink who ate every last egg. Some studies have

the pregnant female crawls out of the water, digs a hole in soft soil with her hind legs, and lays four to eight 1-inch-long elliptical white eggs. She then covers up and pats down the birthing chamber. Some evidence indicates that northern turtles may not breed every year. Short summers limit females' accumulation of fat, which is necessary for yolk production in developing eggs.

The bottle-cap-size "turtlettes" hatch in late summer but instead of emerging into the outside world they winter in the nest, 3 to 4 inches below the surface, where winter temperatures may bottom out at 17° F. Professor Ronald J. Brooks of the University of Guelph discovered the young Painted Turtles' incredible ability to freeze solid and then thaw. Ice fills their abdominal cavity, bladder, and all the spaces outside the cells. Glycerol, a cryoprotectant, or natural antifreeze, prevents the intracellular fluids from freezing, allowing the tiny turtles to emerge healthy and active come spring.

With no help from their parents, they must make their way to the lake. If mom laid her eggs far from the water (as sometimes happens), the task is formidable. Polarized light reflected in the sky above ponds and lakes may guide them to the water but they still have to run the gauntlet of ravens, gulls, and Mink that find tiny turtles a satisfying snack.

Surviving Painted Turtles grow to 3 1/2 inches in four years. Then they become breeding adults able to pass on their genes to a new generation of Western Painted Turtles.

Underwater breathing

Western Painted Turtles can breathe underwater by drawing water in through the nostrils and passing it into the mouth, where oxygen is absorbed.

Sparky says: Help to preserve North America's canoe country wilderness by joining the Friends of the Boundary Waters, 1313 Fifth St. SE, Suite 329, Minneapolis, Minnesota 55414; or the Friends of Quetico Park, Box 1959, Atikokan, ON, Canada P0T 1C0. Both groups work to keep the Boundary Waters and Quetico safe from development.

Wood Frog
Rana sylvatica

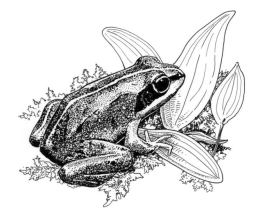

Length 1 ½ to 3 ½ inches

Clay-colored

Distinctive "robber's mask" across face

Spring song is like the quacking of ducks

Eggs and acid

Eggs of the Wood Frog tolerate very acidic waters, even down to pH 4 (a tolerance we can only hope they never have to use).

The Wood Frog truly is a frog of the woods, spending its summers and winters beneath the boreal canopy. The summer canoeist may encounter the masked amphibian in damp, shady woodlands and near tiny streams hunting slugs and bugs.

The pinkie-length (1 3/8 inch to 3 1/4 inch) frog is clay-colored, lacking any green, and wears a black "robber's mask." To complete the masquerade, the Wood Frog can darken or lighten its complexion to match its surroundings.

The earliest breeders in the North Woods, Wood Frogs travel to snowmelt ponds when temperatures reach 50° F. From these temporary pools the males give their curious clacking call, clamoring for females. The sound resembles that of a quacking duck or a hoarse Chihuahua. Opening fishing weekend is a good time to hear the curious calling of the Wood Frog.

Watch for the tennis-ball-size jelly masses of Wood Frog eggs clustered at one end of mating ponds.

Oh . . . and check the antifreeze too!

Wood Frogs freeze solid in the winter, but the freezing is only a temporary condition. University of Minnesota researcher William Schmid discovered the key that unlocked the mystery behind the Wood Frog's strange ability to freeze solid, with heart and lungs actually shut down, and then thaw out in spring to live and jump again. The secret was glycerol, the natural antifreeze, that floods the frog's system as cold temperatures settle upon the northland. Dr. Schmid had discovered the first vertebrate ever known to possess a biological antifreeze.

It works like this. As autumn descends Wood Frogs settle under leaf litter on forest floors. An unknown trigger then shoots the antifreeze process into action. Glycerol is produced very rapidly from carbohydrates stored in the liver, flooding the frog's internal system with sixty times its normal amounts. Humans would quickly lapse into a sugar induced coma and die under a similar dosage, but the Wood Frog's body shuts down. Winter's icy fingers reach the hibernating frog even beneath its insulating leaf and snow blanket, and freeze the critter solid. But glycerol, the biological antifreeze, has done its job, and only the fluids outside the cell walls freeze, causing no permanent damage. In spring, the body thaws and all systems start up. The frog's April journey to the snowmelt breeding pools is fueled by leftover glycerol. Amazing!

Northern exposure

The Wood Frog lives farther north than any other North American amphibian or reptile, surviving above the Arctic Circle in Alaska.

Sparky says: Blindfold a friend with a bandana (the Wood Frog's "robber's mask" is not a blindfold, but it gave me this idea) and lead him or her on a sensory hike in the woods about your campsite. Have your friend feel (moss, tree trunks, holes, crushed cedar, spruce or fir needles), taste (blueberries, Clintonia leaves, raspberries), and hear (wind in the pines, birdsong, hammering woodpeckers). Don't explain what your friend is sensing, just let him or her experience it. Now switch roles and let yourself be blind-led.

Green Frog

Rana clamitans

Length 3 to 4 inches

Males have bright yellow throats

Female's throat is white

Evening song is an unmelodic "tchung"

Fishhook Bay resounds tonight, with a chorus of fifteen or so Green Frogs, all "plunk"-ing away at once, trying to outsing their neighbors. The Green Frog's song is hard to describe. Field guides describe it as a nasal "plunk" or explosive "tchung," or "as if someone has plucked the strings of a badly tuned bass fiddle." These descriptions just don't do it for me, but after a week of contemplation I cannot offer a better one. I imagine some things are better left undescribed. When they hear the Green Frog most travelers believe they are hearing its larger cousin the Bullfrog. Bullfrogs, though, do not live anywhere near the Boundary Waters or Quetico, surviving only as far north as southern Minnesota and southeast Ontario. Their call is a deep three-syllabled "jug-o-rum."

Earlier tonight, at dusk, I stumbled across a male Green Frog who was patrolling his stretch of shore, checking for any intruding males. From a motionless 2 feet away he was a magnificent sight—a dark forest green body with a ridge running down each side of the back and a dazzling sun yellow throat. At 3 inches long, he was near the maximum 4 inches attained by the species. I knew it was a he from the throat color (females have white throats) and the fact that the tympanum, or eardrum, was nearly twice the size of the eye (they are equal size in the females).

Listen for the Green Frogs to begin calling about the third week in June. They often join in concert with the Gray Treefrogs and American Toads.

The shore-lords

If you listen to the evening Green Frog chorus, you'll soon notice that the frogs seem to be fairly evenly spaced along the lake margins. Each one sings from his own well-defined territory, which he defends from other breeding males. Small satellite males are tolerated but must sit lower than the "shore-lord," who often sits higher on a rock or log with his manly yellow throat exposed. The older and larger the frog, the higher he sits.

Singing, chasing, and perching take their toll, however, as breeding males may lose up to 30 percent of their body weight from territory maintenance. As denizens of permanent bodies of water, Green Frogs can afford this expenditure. The familiar early spring frogs—Spring Peeper, Boreal Chorus Frog, Wood Frog—who breed in frenzied territory-less masses in temporal pools that will likely be dry by midsummer, must conserve their energy.

A female chooses a mate with attractive shoreline property suitable for egg laying. She approaches and touches him and then they fall together in amplexus (an amphibious term describing the mating position of the smaller male hugging the female from behind). He gathers thirty to fifty eggs at a time with his hind legs, fertilizes them, and pushes them away. This process is repeated more than one hundred times, until 3500 to 5000 eggs are laid in floating rafts on the water's surface.

Tadpoles hatch in three to five days and remain tailed for twelve to thirteen months, overwintering in the water. They lose their tails the second summer, completing their long two-year immature stage.

Sparky says: Take an evening paddle along a Green Frog–inhabited shoreline. Drift silently and enjoy the chorus, or take a headlamp along to spot the shore-lords on their singing perches.

American Toad

Bufo americanus

Length 3 to 5 inches

Fat and warty

Marked with brown, gray, and reddish brown

Mating call is a long high-pitched trilling

I never want to step on another toad for the rest of my life. It survived, but made a horrific piercing scream that has scarred me permanently. I no longer take carefree night strolls on north country paths without peering into the gloom searching for any wayfaring American Toads. They become primary pedestrians on warm rainy nights, when their favorite midnight snack, the earthworm, comes to the surface en masse.

The American Toad range extends from the Northwest Territory's Great Bear Lake south to Mexico. They live at an altitude of 5,800 feet in Tennessee. Females may reach 5 inches in length, though males rarely exceed 3 inches. Their fat bodies can be brown, gray, or reddish brown after shedding, and sport a lighter racing stripe down the back with large brown warts on either side. Though they're gray, warty, and fat, I still consider them comically cute.

The long steady trill of the American Toad can be heard after the early breeders (Spring Peeper, Boreal Chorus Frog, and Wood Frog) quiet down. They usually begin calling in late May and continue through June.

The trill is created by air drawn into the nostrils and then passed back and forth across the vocal chords. The puffed-out throat serves as a resonator for the sound.

Fifteen thousand eggs in one nest

Five males may battle over one female in their frenzied attempts to mate her and fertilize the five thousand to fifteen thousand eggs she lays in paired jelly strings. The strings, which after absorbing water may reach 72 feet in length and weigh five and a half times as much as the female herself, wrap around sticks and rocks underwater, but stay well camouflaged due to adhering pond gunk.

The black and white eggs hatch in two to twelve days, depending on the water temperature. The warmer the water, the faster the development. Toad tadpoles can be distinguished from frog tadpoles by their jet black color and communal groupings. They are weak swimmers, but after a month and a half to two months the tailless minitoads storm the beach. Maturity comes at age three, but cautious toads can expect a long life. One European Toad lived to the rocking-chair ripe old age of thirty-six.

After a lazy summer of worm and slug hunting it's time to settle in for the long winter sleep. Toads find a patch of loose soil and start digging. Kicking dirt with their hind legs, they corkscrew down into the dirt until they are below frost line, which may mean 6 or 7 feet in the canoe country.

In the spring the whole cycle starts anew.

Kiss at your own risk

Of course toads can't give you warts, but the "warts" behind their eyes do contain poison glands. This poison, known as bufogin, prevents dogs from ever trying seconds of toad. Raccoons are smart, though, and avoid the poison glands by flipping the warty beasts over to attack the belly. Skunks and Eastern Hognose Snakes are apparently immune to bufogin's effects. Humanoids should wash their hands after handling toads to prevent passing the irritating substance to the eyes.

Chinese medicine

Toad skin is a medicine in China. It contains adrenalin and is used to raise blood pressure.

Sparky says: Time the length of American Toad trills during different temperatures. Do toad calls last longer during warm evenings or cool ones?

Amphibians and Reptiles of the Boundary Waters and Quetico

AMPHIBIANS

Order Caudata: Salamanders
- ❏ Mudpuppy — *Necturus maculosus*
- ❏ Spotted Salamander — *Ambystoma maculatum*
- ❏ Blue-spotted Salamander — *Ambystoma laterale*
- ❏ Red-spotted Newt — *Notophthalmus viridescens*
- ❏ Eastern Red-backed Salamander — *Plethodon cinereus*

Order Salentia: Frogs and Toads
- ❏ American Toad — *Bufo americanus*
- ❏ Spring Peeper — *Hyla crucifer*
- ❏ Gray Treefrog — *Hyla versicolor*
- ❏ Boreal Chorus Frog — *Pseudacris triseriata*
- ❏ Wood Frog — *Rana sylvatica*
- ❏ Northern Leopard Frog — *Rana pipiens*
- ❏ Green Frog — *Rana clamitans*
- ❏ Mink Frog — *Rana septentrionalis*

REPTILES

Order Testudines: Turtles
- ❏ Snapping Turtle — *Chelydra serpentina*
- ❏ Western Painted Turtle — *Chrysemys picta*

Order Squamata: Lizards and Snakes
- ❏ Northern Red-bellied Snake — *Storeria occipitomaculata*
- ❏ Eastern Garter Snake — *Thamnophis sirtalis*
- ❏ Northern Ringneck Snake — *Diadophis punctatus*

Insects and Other Invertebrates

Mayfly
Order Ephemeroptera

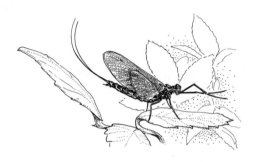

Flutter awkwardly on clear, delicate wings with two long filamentous "tails" trailing behind

Found near water in June

More than 550 species of mayflies inhabit North America. Delicate wings held straight up over the back and two long filamentous "tails" characterize this insect order.

The hatch

We are all familiar with the mass emergence of mayflies. Television has made it one of their annual rites of spring. Every May we see footage of some town along the banks of the Mississippi River alive with fluttering mayflies. Then they cut to bridge scenes, where the road is buried a foot deep in dying mayflies. Cars slip and slide on the squashed bodies. Even the snowplows are brought out.

Did you know the same miracle of nature's timeclock happens every June throughout the canoe country? It does, but on a lesser scale.

The adults have to make the most of every second, since their above-water life usually lasts only a day or two. Eating is not a concern since their mouth parts are useless. Energized only by food taken in during their underwater nymphal stage, most get down to the nitty-gritty and mate on the wing. The male flies up and seizes a likely female. Touch and chemical signals tell them whether they are of the same species. The female then drops five hundred to one thousand eggs into the river or lake from which she came just minutes or hours before.

Mayfly eggs possess remarkable mechanisms that aid in their survival. Most mayfly species lay eggs with sticky coverings, which help anchor the egg underwater. In certain species, when the egg touches the water, fine threads uncoil. Some threads have adhesive disks that anchor the egg to the lake bottom. Mayflies take from six weeks to three years to develop into adults, passing through upwards of twenty-one molts as aquatic nymphs.

The benefits of the mass emergence are clear: mayflies have their best chance to meet others of their kind. The benefits for trout, bass, perch, Walleye, and birds are also obvious.

Sparky says: Make your own "web of life." All you need is a ball of string and your group standing in a circle. The "mayfly person" holds the end of the string. What eats mayflies? The person who answers "trout" is connected to the mayfly by passing on the ball of string. The string passes to the person answering the question "Who eats trout?" What if the Osprey who ate the trout dies? Beetles may eat on the carcass. Woodpeckers come to snatch up the beetles. Woodpeckers nest in trees, which can support the nest of Bald-faced Hornets. Hornets capture flies to feed their larva. Male mayflies feed on flowers. Flowers have roots in the soil. Seeds in the soil are watered by rain. Trees grow from the earth. They give off oxygen, which humans breathe. We cut the tree down to burn in our stoves to keep us warm. The remaining stump rots providing other seeds with a seedbed and nutrients. Connect each person who answers, with the string, until you have a "web." Push on the web. Does everyone feel it? Everything is connected in an ecosystem. What if all the "plant people" let go of the string? The web falls apart just like the ecosystem would do if it had no plants. Try the "web of life" on your next canoe trip. It is a powerful learning activity.

White-tailed Dragonfly

Plathemis lydia

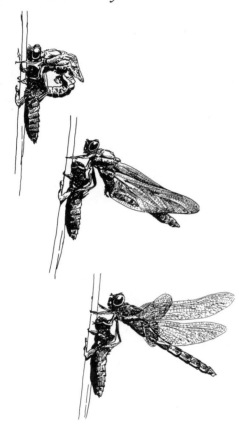

Length 1 to 3 1/2 inches

Two pairs of veined wings held straight out at sides when at rest

Often seen at midday, patrolling lakeshores for small flying insects

I've always wanted a pet dragonfly. I thought I'd tether it to me by a long thread so it could fly freely, capturing all those nasty mosquitoes, blackflies, and deerflies that buzz my head on a long portage. Come to think of it, I might need a whole team of them.

Dragonflies are exclusively carnivorous as both nymphs and adults. Bulging compound eyes, composed of thousands of facets, help them pinpoint any micromovement made by a nearby bug. Two pairs of veined wings powered by incredible flight muscles allow dragonflies to zip around, stop on a dime, hover, back up, and zip off again at speeds of up to 30 miles per hour. Dragonflies are efficient insect-killing machines. And they are totally, completely, entirely harmless to humans.

Canoe country dragonflies can be divided into three groups: darners, clubtails, and skimmers.

Darners: family Aeshnidae; large, 2 1/2 to 3 1/2 inches; metallic-colored, mainly browns, blues, and greens; found near ponds and swamps; sexes are the same color; compound eyes meet and the hind margin is nearly straight; they are swift and high flying.

Skimmers: family Libellulidae; 1 to 2 1/2 inches long with wingspan noticeably greater than body length; many with spots or bands on wings; found near ponds and swamps; sexually dimorphic; compound eyes meet but the hind margin is deeply lobed; fast flight interrupted by hovering.

Clubtails: family Gomphidae; most species are 2 to 3 inches and drab-colored; abdomen is often swollen at the tip to form the "club"; found along streams and shores of large lakes; sexes are the same color; compound eyes do not meet on top of head; they fly steady without periods of hovering, some undulate in flight.

With patience and practice you can easily identify these three dragonfly families. Warm, sunny, and still makes the best weather for observing them. Dog days are dragonfly days.

The wheel of mating

The male White-tailed Dragonfly is a skimmer with a stubby bluish white abdomen. Females have a dark abdomen. As with most skimmers, its body length is noticeably shorter than its wingspan. The territorial males each claim a piece of shoreline property. When a receptive female happens by, the dominant male in that territory grabs her and tips his abdomen over his head as tiny pincers clasp her behind the head. She responds by lifting her abdomen tip to his second abdomen segment where the sperm packet awaits transfer. Mating lasts only ten seconds. This position, called the "mating wheel," is performed while hovering in midair. But the successful male must be on guard, for the female stores sperm, and the last male to mate with her will pass on the most genes. Mate guarding is the strategy. As she looks for a spot to lay her eggs, he hovers near her, preventing any last-minute interlopers from zipping in and mating with her. Some skimmers even go so far as to "contact guard," the male staying attached to the female even while she deposits eggs. She lays eggs by dipping her abdomen into the water. The eggs settle to the bottom and hatch.

Giant dragons

Dragonflies with 2-foot wingspans once cruised between giant ferns of the Pleistocene. All that remains of these extinct species are fossils.

Aliens!

Dragonfly nymphs are voracious underwater predators. They possess a hideous double-hinged jaw armed with sharp, jagged teeth that can be instantaneously extended to grab prey. Even prey as large as a minnow, which is a whale to a half-inch-long nymph, can be taken.

Nymphs breathe by extracting oxygen from water drawn in through their abdomens. The expulsion of the waste water helps propel them swiftly about. The White-tailed Dragonfly nymph undergoes ten-plus molts before emerging onto dry land after a winter underwater.

The nymph crawls inland, then 10 to 20 feet up a tree before the transformation. Then, as in a scene out of *Aliens*, the back of the exoskeleton splits open and an adult dragonfly emerges, wings and all. I witnessed one such remarkable event. It took the dragonfly twenty minutes to emerge and another thirty minutes for the wings to be pumped full of fluids. Only a hollow, papery exoskeleton, like so much excess packaging, clings to the tree once the adult flies off.

 **Sparky says:** On a sunny, still day go down to the lake edge to watch the interaction of male White-tailed Dragonflies. Each male defends his own piece of turf, which he patrols all day. Watch for the following three behaviors.

Perching—A male perches on logs, rocks, plants, and sticks, above and facing the lake, watching for territory-invading males and receptive females.

Chasing—If another male enters the territory, the territorial male will fly directly at him to chase him off, the two will face each other while hovering, then fly up in the air several yards. The territorial male will fly under the intruder and bump him from underneath. You can hear the wings contacting each other. They both then fly way up in the air and off to one side.

Patrolling—The territorial male flies in a slow circuitous route around his territory, eventually returning to his perch.

Some territories have a dominant male with one or two submissive Whitetails. Watch for the raised abdomen (about 75°) of the dominant one when he encounters the abdomen-dipping underlings.

Binoculars will greatly aid your viewing. So sit back, get a tan, and enjoy the show.

Damselflies hold their wings vertically over their backs when at rest. This differentiates them from dragonflies, who hold their resting wings straight out to the sides. Slower wingbeats also make damselfly flight look fluttery, almost butterflylike, compared to the direct zipping flight of dragonflies. Damselflies are dainty and slender.

Shady streams are the favored haunts of the Black-winged Damselfly. Truly a stunning creation, the male's body is a shimmering metallic green, as gaudy as a 1960's smoking jacket. His wings are smoky black. The female sports a brown body with tiny white spots at the front edge of the black-tipped wings. Midmorning to late afternoon on warm, calm days are the best times to observe their activities.

Cross River congregation

Tubing (and beaver-dam hopping) down the Cross River, which flows between Ham Lake and Gunflint Lake on the east side of the BWCAW, on a hot, sunny, lazy July day, I watched hundreds of Black-winged Damselflies flutter along both banks of the stream. Their irridescent green bodies shimmered in the sunlight.

I noted in my journal that they were "chasing and confronting one another." This piqued my curiosity. So later that month I floated on down to the library. I have always been amazed at the amount and type of information stored in that one building. And again I was not to be disappointed. Fluttering over to the 595s (Insects), I found not only illustrations of the species but references in many books to its behavior, including one book that contained an entire chapter on the courtship behavior of the Black-winged Damselfly!

I discovered that the Blackwings are not exclusively a stream-edge species but congregate there to find mates and lay eggs. The male defends a "miniterritory" that must include some aquatic vegetation for the female to lay eggs on and a perch with a commanding view of his domain. This defending

Black-winged Damselfly
Calopteryx maculatum

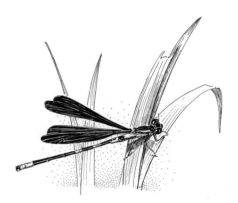

Two-inch-long, slender bodies

Males are metallic green with smoky black wings

Females are brown with tiny white spots at the front edge of their black-tipped wings

Wings held straight over back when at rest

activity is most likely what I noted in my journal. The males exhibit several displays, including the following.

Wing spreading—When another male enters a Blackwing's territory, the protector spreads his wings horizontally and raises his abdomen. If this doesn't work, a chase ensues.

Cross-display—If a female happens by, the perched male spreads his back wings and keeps the forewings held high over his back. The abdomen is raised. To further impress her, he lifts off and flutters rapidly while facing his prospective mate.

The mating position in damselflies, as in dragonflies, is called the mating wheel. The male transfers to the female his sperm packet. But he must stay on guard, because another male can zip in and deposit his own sperm packet, thereby passing on his genes instead.

Couples honeymoon in the male's territory. She deposits the fertilized eggs in stems of aquatic plants just below the surface while he continues to jealously guard her. After she finishes depositing the eggs, she flies off, leaving him to mate with other females.

Sparky says: Keep a journal of your next canoe trip. Record daily events, route notes, personal reflections, and impressions of specific lakes. Also keep tabs on the natural world by jotting down the behavior of animals you observed, questions to be looked up later, lists of species seen, and sketches. Remember to record temperature, wind, cloud cover, and precipitation for each day. A little bit of work reaps great rewards each time you reread your trip journal.

If you've ever seen a white frothy mass of "spit" on the stem of a plant along a portage or at a campsite, you have encountered the natural habitat of the spittlebug (froghopper nymph). A spittlebug or two hide beneath the protective, superficially repulsive, froth. Gently part the froth to reveal the tiny creatures.

Spittlebugs are about $1/8$ inch long, light green, and oval shaped. Nymphs go through several molts as they live within the froth, changing in size and color. Adults, now called froghoppers, leave the security of the spit, preferring a "hopping" life amongst the foliage. They can reach $1/4$ to $1/2$ inch. Their wide-bottomed shape makes them look like micro-frogs, hence the name.

"Snake spit" and other myths

It must be too simple to bend over, reach down, and investigate the blobs of froth. Or maybe the resemblance to spit disgusts people and deadens their curiosity. Whatever it is, people for centuries have made up bizarre tales of "frog spit," "cuckoo spit," and "snake spit" to explain the froth blobs.

The true explanation is much more interesting. Come fall, the adult female lays her eggs in a food plant stem. The eggs overwinter and hatch into nymphs, which develop into adults on the stem of the plant, exposed for all the hungry world to see. Security comes in the form of froth, which is a by-product of the nymphs' waste.

Nymphs suck plant juices for nutrition and slowly excrete excess juices. This waste fluid contains an enzyme that breaks down a wax that is produced simultaneously by the anal glands, resulting in a mixture nearly unique in the animal world—a liquid waxy soap. The nymph dips its abdomen into the liquid and essentially starts blowing tiny bubbles. Little bursts of air form one bubble at a time, or about one every second, until the nymph is totally covered. Home, sweet, home. To breathe inside the froth, the spittle bug simply sticks its abdomen into a

Spittlebug (Froghopper)
Family Cercopidae

Nymphs (Spittlebugs)

Average $1/8$ to $1/4$ inches long

Found in frothy masses of "spit" on plant stems

Adults (Froghoppers)

Average $1/4$ to $1/2$ inches long

Frog-shaped

Hop amongst foliage

single bubble, allowing air to be channeled to spiracles on either side of the body.

Protection is the major benefit of this behavior. Birds and small mammals may have a real aversion to sticking a beak or snout into the mass and mucking about for a paltry meal. Some predatory insects are undaunted though. Damsel bugs suck out the body juices of the nymphs through the froth, while digger wasps pull the nymphs out whole.

The froth may also prevent desiccation, aid in respiration, provide a constant climate, and help prevent parasites.

Sparky says: Find a blob of froth, crouch down, and gently part it to reveal the spittlebug. They don't bite, so let one crawl on your finger. Observe it. Return the nymph to the frothy area. If you are patient, sit and wait for the nymph to pierce the plant stem and begin feeding again. Watch the nymph make more bubbles.

C anoeists share the waterways with many critters. One of the paddler's closest companions is the water strider, which skitters over the water, evading the oncoming bow. To a creature as long as your fingernail is wide, the bow of a canoe must be a horrific sight, as if the *Queen Mary* were bearing down on you. Fortunately, water striders are wonderfully created for short bursts of speed across the water's surface. Yes, they run on top of the water, supported only by surface tension.

Watch for water striders along the edges of lakes and rivers. They require quiet waters for prey detection and mating, so they never stray far from a sheltering cove. Water striders can be seen from early spring, when the adults emerge from an on-land hibernation, through late fall.

Walking on water

Water striders' remarkable ability to walk on water has a logical explanation. The tips of their wide-spread four hind legs are fringed with water-repellant hairs that keep them atop the water's surface film. If a wavelet should happen to wet one of the water strider's legs, it must crawl out of the water and dry off to regain its buoyancy. On sunny days in shallow sandy bays, an interesting phenomenon occurs that enables you to actually see the depressions water striders leave on the surface film. Shadows are cast on the bottom, showing striders with "clown feet," which are really shadows of the surface-tension depressions created by the feet.

Sensory organs in the water strider's legs enable it to feel vibrations on the water's surface when a "victim" falls into its pool. Quickly locating the fly, mosquito, or moth, the water strider grabs it with shortened grasping forelegs and pierces it. The water strider injects the victim with a juice that dissolves its innards, creating a tasty "fly shake" or "moth malt" ready to be sucked out.

Water striders create their own ripples to communicate

Water Strider
Family Gerridae

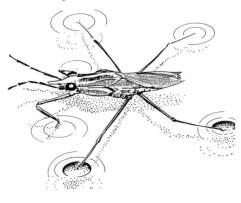

Long-legged aquatic insect that walks on water

Found in groups along quiet shoreline waters

Trout finders

Water striders foraging exclusively along stream banks indicates that trout are present. Troutless streams have striders bravely foraging in the center of the channel.

messages of love and anger. The male hangs out near a good egg-laying site and taps out a love message with his middle pair of legs for all the neighborhood females to hear. He sends out a different signal to keep the other males away. Unfortunately, at ten to thirty ripples per second, these signals are too fast for us to see.

Sparky says: Watch for the strange shadows created by surface-skimming water striders on shallow sandy lake bottoms. Also, try dropping a stunned mosquito or fly onto the water near a bunch of striders. Do they grab it right away? How do they eat it?

On one tree trunk sat four of the largest sawyer beetles I had ever seen. Their ridiculously long 3-inch antennae curved out and away from their heads in graceful arcs. Their thick rectangular bodies were a third the antenna length and grizzled gray.

The two pairs were in a strange position. The forward one had its jaws firmly locked into the tree bark. The rear beetle was off to the side with one leg up on the back of its partner as if encouraging him or her to bite into the bark harder. I've yet to research this behavior but am almost positive it has something to do with reproduction.

Sawyer beetles are members of the family Cerambycidae—the long-horned beetles—so named for their long antennae, which are at least twice as long as their bodies. A rear-slanting face and curved antennae give their faces a ramlike appearance. Most canoe country sawyer beetles are 3/4 inch long with 2-inch antennae and are speckled gray—the color of Balsam fir bark. It seems that the larger the sawyer beetle the lighter its body color. The stout and powerful jaws present an effective defense but also perform the delicate task of feeding on pollen and flower parts. The sawyer would rather escape than fight. If disturbed while resting on a tree trunk, a sawyer would most likely release its grip on the bark and fall to the ground, escaping beneath the leaf litter. Flight is a less desirable escape option because they fly slow and direct, much like a traffic helicopter hovering above a crowded interstate.

Sawyer Beetle
Monochamus species

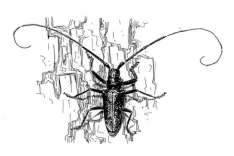

3/4-inch-long rectangular beetle with curved antennae twice as long as body

Bark-colored

Faces have a ramlike appearance

Larvae bore into downed trees and make audible chewing noises

The squeaky rocking chair

Female sawyer beetles lay their eggs in bark crevices on downed trees. The eggs hatch into whitish larvae that feed on the inner dead wood and grow into thick cylindrical fleshy grubs. The "creeeak . . . creeeak . . . creeeak" sound, like that of a squeaky rocking chair, is made by the grubs as they bore deep into the wood. Amazingly loud, the chewing noise can be heard from many yards away.

A U-shaped pupal cell is formed by the larvae deep in the log and is plugged with wood frass (insect droppings). In the security of the chamber the grub pupates into an adult which then chews through the frass and emerges into the outside world leaving behind only a perfectly round hole about 1/4 inch in diameter.

 Sparky says: Shhh! Listen carefully. Do you hear it? Its the "creeeak . . . creeeak . . . creeak" of a sawyer larva chewing dead wood within a fallen log. Now it's time for a Mission Impossible task. Your assignment, if you should choose to accept it, is to get close to the "chewer" and its log. Many times I have tried and failed. If you can actually touch the log while the grub is still "creaking" you are a first class special agent. This page will self destruct in five seconds.

Whirligig Beetle

Family Gyrinidae

The common name and family name of this abundant aquatic beetle should clue you in to its diagnostic behavior. "Whirl," as in whirligig, and "gyre," as in Gyrinidae, are synonyms for a spiral turn. Gyre, as in gyrate and gyroscope, appropriately labels this black, pinkie-fingernail-sized oval beetle that makes mesmerizing circular patterns on the surface of northern waters in search of food. Gregarious by nature, they form large groups (flocks? herds?) in quiet water. They are common on Boundary Waters and Quetico lakes from spring through fall, with a lull in early summer, the period after last year's adults lay eggs and die and before the new generation of whirligigs grows up to "whirl" on their own. Adults overwinter in mud at the water's bottom.

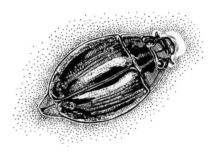

Small oval black beetles

Spin and whirl in groups on the water's surface

Old four-eyes

You would be somewhat correct if you yelled "Hey, watch out, four-eyes!" at the whirligig beetles scurrying from the path of your oncoming canoe. The whirligig doesn't actually have four discrete eyes; it has one pair of eyes split into two halves, each with a specific function. The "pair" below the waterline watches for predators ("beetle-eaters") while the other "pair" scans the surface above the waterline for the beetle's own lunch.

To aid in finding lunch, whirligigs possess sensory organs (called Johnston's organs) at the base of their short antennae, which lay flat on the water and pick up the distinct ripple vibrations of a struggling fly, mosquito, or moth. Johnston's organs allow the whirligig to easily distinguish "victim ripples" from "whirligig ripples."

Four flattened hind legs act as paddles, propelling whirligigs in their circular watertop foraging patterns. The front pair are the grasping legs for catching and holding prey.

Scuba divers

Whirligig beetles breathe through spiracles, openings in their

173

body that deliver oxygen to all the necessary parts. But when they dive underwater, atmospheric oxygen is not available to them. Humans encounter the same problem when underwater. Whirligigs have solved the problem by carrying with them an air bubble on their hind end whenever they dive, just as scuba divers carry their own oxygen supply. Spiracles open up into this bubble and readily take in the available oxygen. Whirligigs can also replenish the bubble's store of oxygen once it has been used up. When the "tank" is nearly empty, the partial pressure of oxygen in the bubble is less than that in the surrounding water, so oxygen from the water diffuses into the bubble, replenishing the tank with clean breathing air. Since nitrogen in the bubble diffuses into the water at a much slower rate, the bubble retains its shape and remains functional for perhaps twenty minutes.

 Sparky says: Challenge your group to figure out why whirligig beetles have also been called "submarine chasers," "lucky bugs," and "write-my-names." Compare and share your answers. Make up your own name for these whirling, spinning, diving, bubble-breathing beetles.

Fireflies electrify the warm, still nights of summer. Their bioluminescent flashings can ignite even the most dormant of imaginations. Watch for them in June and July. Lie on your back and enjoy the show.

The firefly, or lightning bug, is neither a fly nor a true bug; it is a soft-bodied beetle of the family Lampyridae ("light-producer"). Most noticeable at night, they are often overlooked during the day as they rest on low vegetation. Canoe country visitors just don't recognize them for what they are. Adults are 1/2 inch long and oblong. A Darth Vader–like hood, called a pronotum, conceals their head. They range from dark gray to black and are marked with orange and yellow. Legs and antennae are nearly the same length.

More than two thousand species of fireflies inhabit this planet, and a number of them flash their hind ends in the canoe country. Fireflies use their glowing Morse code primarily for attracting a mate of the same species, so each species speaks its own love language. Mating signals vary in duration, interval between flashes, number of flashes, distance flown between flashes, and color of flashes. Different species also signal at different times of the evening, at different times of the season, and from different habitats. Males flash while flying and females flash from a stationary position on vegetation.

Fireflies of different genera emit different colors of light. *Photinus* fireflies glow yellow; *Photuris* fireflies glow green; *Pyractomena* fireflies glow orange. With practice you can differentiate the unique signals quite easily.

A light in the forest

Greek philosopher Aristotle, more than two thousand years ago, observed that "some things though they are not in their nature fire nor any species of fire, yet seem to produce light." Many myths arose to explain this mysterious light in the night. But it was not until 1887 that French scientist Raphaël Dubois unlocked the mystery of bioluminescence. He discovered that

Firefly
Family Lampyridae

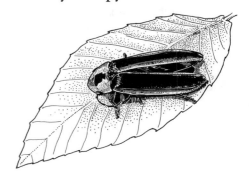

Soft-shelled oblong beetles

Length 1/2 inch

Dark gray to black, often marked with orange and yellow

Most conspicuous at night when their hind ends flash green, orange, or yellow

Living light bulb

It would take twenty-five thousand fireflies in a jar, flashing in unison, to equal the light intensity of one 60 watt bulb.

Fooled

Sixteenth century English privateer Sir Thomas Cavendish and his men fled the West Indies to waiting ships when a sentry mistook firefly flashes ("infinite number of moving lights") for a Spanish attack.

when the pigment luciferin is combined with the enzyme luciferase in the presence of oxygen, light is produced. In other words, the oxidation of luciferin produces light.

And what an amazing light it is: a "cold light," which means that nearly no energy from the reaction is lost to heat. An unbelievable 98 percent of the reaction energy is given off as light. In contrast, the average household light bulb is only 10 percent efficient, with 90 percent of its energy lost to heat. Science has attempted to duplicate the efficient cold light of bioluminescence, but it has proven far too expensive.

Fireflies are masters at controlling their own light production to create unique flash signals. When they open their tiny abdominal pores, oxygen flows in and instantaneously oxidizes the luciferin. This creates light and their abdomens glow. They close the pores and the light goes out.

Mating by Morse code

The male flys about in the darkness of evening, casting forth his luminescent love signal, anxiously awaiting the exact duplicate signal from a leaf-sitting or grass-clinging female below. They trade love signals until they home in on one another and mate. Only then does the abdominal signaling cease. But firefly, beware! Females of the genus *Photuris* are out there just waiting to deceive males of another genus. She can mimic the flash signals of up to seven other firefly species, enabling her to lure males of different species to her. Before they discover they've been duped, she eats them.

Sparky says: Be a firefly flash detective. First tell the males from the females. Males fly-and-flash; females sit-and-signal. Once you've mastered that, try to tell the three genera apart by their flash color: green, orange, or yellow. If you want to be a master detective, try recording the flash signals. Remember the several different species of firefly in the canoe country each have a unique pattern of flashes.

Here are some examples:

Photinus obscurellus—two or three quick yellow pulses, six-second pause, two or three quick yellow pulses

Pyractomena angulata—eight to ten flickering orange flashes

Photuris pennsylvanica—bright green flash, pause, sustained green glow

How many different flash signals did you detect?

Kids might enjoy creating their own flash signals. After dark, have them stand 100 feet or so apart, each with a flashlight. One flicks the light on and off in a certain pattern, which the other tries to imitate.

A healthy glow

A case has been documented of a frog that ate so many fireflies that their glow could be seen through the skin of its belly.

Luna Moth

Actias luna

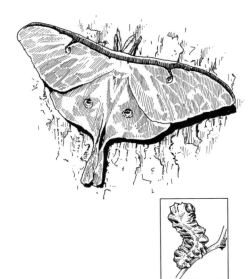

Large moths with long, delicate tails

Pale green wings span 5 inches

Nocturnal

Most common in June

Five Luna Moths sleep just outside my cabin door as I write this. I go out to admire them frequently, hoping to be inspired. They are large and pale green, with long, delicate, twisted tails. Their color, one not normally found in nature, is an undescribably beautiful mint green that goes so well with its maroon and yellow highlights. During the day, the nocturnal Luna Moth hangs in the green foliage of deciduous trees and looks for all the world like just another leaf—which is the whole idea—to camouflage itself from predators while it gets some shut-eye. By pure chance, I once found a mating pair hanging from a Mountain Maple at waist height.

How did I get five Lunas to come to me? Well, I left the porch light on last night and attracted not only the Lunas but two Polyphemus Moths and twenty others of assorted species. Keep your eyes open as you grunt down the portage path, and you too may encounter the most spectacular of our giant silkworm moths. Luna Moths are most commonly seen from late May to late June.

The naming of this moth has a planetary theme. It belongs to the family Saturniidae, is nicknamed the "moon moth," and has Luna (pertaining to the moon) as its common and Latin species name. This moth is out of this world. It has a wing-spread of 5 inches and a long dangling tail. It appears to be made of "pale green gauze," North Woods author Helen Hoover once wrote. The front margin of the wings is maroon, as are the legs and eyes. The body is white and furry. Males have large feathery yellow antennae, while females have narrower ones and a proportionately larger body. Two transparent, narrow-edged "eyes" dot the forewings, which the moth holds out horizontally from the body when at rest.

Living batteries

Luna Moths, simply put, are living batteries running on birch and aspen leaf energy stored by the mighty munching caterpillar the summer before. In fact, the adult moths don't even have mouths or stomachs. Males are simply breeding machines meant to mate and die, while females have the added responsibility of laying eggs. The battery usually runs down in less than a week.

By then the two hundred or so white eggs have already been laid and cemented to the twigs and leaves of its food plant—birch, aspen, or willow. The green leaves feed the green caterpillars that emerge three weeks later. All summer the caterpillars fatten up, and in the fall they spin cocoons of their abundant silk among the leaf litter where they will spend the winter. Come late May in the canoe country, the metamorphosed moths secrete a fluid that dissolves the tough silk cocoon, allowing them to emerge in all their resplendent glory with batteries fully charged.

Sparky says: Try baiting moths. If you are staying in a cabin, leave the porch light on all night. It will attract many beautiful night-flying moths. You can also observe them while they sleep during the day.

If you're on a canoe trip, try baiting them with smell. Mix up two rotten pears, a few black bananas, a whole 2-pound bag of brown sugar and two or three cans of cheap beer and put it in a tight-fitting plastic container. (Prepare the mix before your trip, since cans are banned from the BWCAW.) Smear the mixture on the bark of several trees a good distance from your campsite. Check each tree with a flashlight at dusk and dawn for any moths (or bears) you might have attracted.

Bat and moth

A study in Leeds County, Ontario, by J. E. Yack, uncovered an interesting correlation between the timing of Luna Moth activity and bat activity. It seems that Luna Moths are most active in May and June, while bats that feed on moths peak in activity in July and August. Early-season emergence helps the Luna avoid one major predator.

Tiger Swallowtail

Papilio glaucus

Yellow with black tiger-stripes

Wingspan up to 5 inches with 1/2 inch tails

In early June large concentrations may be seen on sandy shores

After spending the long cold winter snug in the security of its tough papery cocoon, the brilliant yellow butterfly with black tiger-stripes emerges in full glory. Beginning at the end of May, and well into June, the paddler may encounter groups of Tiger Swallowtails on wet sandy shores drinking to replenish liquids lost in emergence. More often you'll encounter them singly, flying about or feeding on flower nectar. Don't look for them on cold cloudy days because they need a body temperature of 82° to 105° F for controlled flight. They achieve this by basking, wings held open to collect the sun's rays—real solar-powered bathing beauties.

The Tiger Swallowtail has a wingspan of up to 5 inches with 1/2 inch tails trailing from the hind wings. These are their "swallow" tails. All summer and into early fall the Tiger Swallowtail may be seen in the canoe country.

If you are super-observant you may spot their tiny round green eggs laid on the top of plant leaves such as birch, aspen, willow, cherry, and Mountain Ash. Ten days after the females lay eggs, caterpillars hatch and immediately start munching. The smooth green larva grows to a full 2 inches, with a swelled front end and two "eye" spots. When a caterpillar is disturbed, it pushes out a pulsating orange scent gland that gives off a repugnant odor. The overall grotesque effect is a vital protective masquerade.

The boys at the bar

There they are again, the same place I left them four years and eight days ago, a living, shimmering, vibrant mass of yellow. The mass, upon closer inspection, reveals twenty to thirty Tiger Swallowtails feverishly taking up water from the wet sand through their delicate but efficient threadlike proboscises. The mid-June noon sun overhead warms the sand grains of Ham Lake in Minnesota's Superior National Forest. The swallowtails remain oblivious to me as my face inches toward them. My eyes are now less than 6 inches away from the quivering yellow-

and-black tiger-striped wings. (Come to think of it, I've never seen a yellow-and-black tiger. I've always thought of them as more orangish. Oh well.) I lift my glasses to allow my myopic eyes the full freedom to do what they do best—focus close. Scales shimmer. I notice that all the swallowtails have claspers on their hind ends, which means they must be males. Revelation! I had stumbled upon the Tiger Swallowtails' all-male drinking club.

Males emerge several days before the females and head right for the beach to have a drink with their friends and wait for future mates. Entomologists believe that this grouping facilitates mating, making it easier for the later-emerging females to find a mate. The males will also be dried, exercised, hydrated, and in full potency to greet the females.

Swallowtails gather for other reasons too. Two hundred were counted swarming a farmyard manure pile, presumably taking in essential nitrogen and sodium and possibly introducing beneficial bacteria into the gut. I have seen White Admirals in the BWCAW gathering on gull guano for the same purpose—butterfly salt licks. Tiger Swallowtails also sleep together in night assemblies.

Sparky says: Try a game of animal charades on your next layover day. Choose up sides, make a list, and let your creative juices flow. Act out the animal's behavior or its name while your team struggles to guess. How would you act out "Tiger Swallowtail"?

Tobacco smoke

Tiger Swallowtails are attracted to tobacco smoke.

Mosquito
Family Culicidae

Delicate fly, 1/2 inch long or less

Long legs and thread-like piercing beak

High pitched whining noise made by wings

Abundant late May to mid-August

Emerge at dawn and dusk

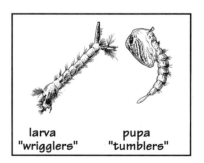

larva	pupa
"wrigglers"	"tumblers"

Ahh, mosquitoes . . . the scourge of the North! Or was that blackflies? Or deerflies? Oh well, they share the title as far as I'm concerned.

I will skip the paragraph of text that tells the reader where to find this animal, since one word will suffice . . . everywhere. I can tell you how to avoid the critter. Plan a trip for the period from ice-out until May 15, or the period from August 15 until ice-up, or during winter. But mosquitoes truly are a part of the whole Boundary Waters–Quetico experience. I've always said "the best bug dope is a positive attitude," and it's true. Also if you camp on high, exposed, windswept campsites your problems will be few. The one, and most crucial, thing to remember for preventive mosquito madness management is to always . . . always . . . have your firewood gathered, tent up, supper made and eaten, dishes done, and teeth brushed before dusk settles in and the buzzing, bloodthirsty, winged vampires invade.

It may surprise some people to learn that there are more than 50 species of mosquitoes in Minnesota alone. There are 165 North American species and 2700 species worldwide. All mosquitoes are true flies in the order Diptera.

Males have bushy antennae and don't bite. They feed on flower nectar and do not need a blood meal for egg development as females do. Most are 1/2 inch long or less.

If you see what appears to be a humongous mosquito with legs spanning 2 inches or more, you've encountered the harmless, nonbiting crane fly, a cousin to the mosquito.

Life Cycle

Let's start with egg laying. After mating, and oftentimes after a blood meal, the female lays eggs in pools of stagnant water. *Culex* genus mosquitoes lay "rafts" of up to several hundred eggs on the water's surface, while *Aedes* eggs drift singly to the bottom. Eggs can hatch in one to five days, or they may survive years of desiccation till rain again creates a life-giving pool.

The eggs hatch into larval "wrigglers." Wrigglers hang

upside down from the water's surface, breathing through an abdominal "straw" that breaks through the surface tension. The Cattail Mosquito wriggler uses a snorkel to breathe. It pierces submerged plants to get at oxygen stored in plant cells. Wrigglers chow on the microscopic organic matter known as plankton.

Wrigglers transform into pupae, called tumblers, after a week or two. When disturbed, tumblers tumble downward. An active pupal stage such as this is rare in the insect world. In two to fourteen days, depending on species and water temperature, adults break the pupal case and fly off. I witnessed such a hatch in stagnant pools on the rocks of Artist's Point in Grand Marais. The adult mosquitoes just seemed to materialize from the pool. Try as I might to actually see one emerge, I could not. Empty pupal cases littered the pool's surface.

Adults may live minutes, hours, days, weeks, and even up to three months on the wing. The House Mosquito actually hibernates through the deep dark winter.

Blood Lust

Males search for flowers to feed on while females are out for blood! Females need protein to form their eggs, and protein is found in blood. But not just any blood will do. Mosquitoes specialize in their blood quest; some attack only mammals, others birds, while yet others prefer frog blood. Victim specialization may even extend to subgroups within these categories. When the female finds a suitable host she lands and bites. But a mosquito does not just jab its mouthparts into the skin. First it cuts a minute hole with its razor-sharp micoscopic stylets to allow the strawlike proboscis, or feeding straw, access to the blood. At the same time, the mosquito secretes an anesthetic so the bite won't hurt as much and an anticoagulant so the blood won't clot and clog the mosquito's proboscis. Then the pump starts. A full blood cargo will double the mosquito's weight and severely impede its flying ability. Swelling of the gut during

Death pool survivor

The Bog Mosquito lays its eggs in the acidic "death pools" inside the specialized leaves of the Pitcher Plant. The water dissolves other trapped insects but is harmless to the developing Bog Mosquitoes.

Orchids

Male mosquitoes are believed to pollinate certain bog orchids in the Canadian Arctic.

Travel

Certain mosquito species may travel 50 to 70 miles in their two-to-three-month lifetime.

blood intake presses on a nerve that triggers hormonal secretions necessary for egg maturation. In some mosquito species, the female produces only a single egg if she finds no blood meal, while successful suckers take in enough blood to make seventy-five to five hundred eggs, depending on the species.

Mosquitoes and humans

How do they find us? Carbon dioxide that we exhale forms a scent trail for the mosquito to follow. Once close, they can detect infrared radiation (heat) coming from exposed body parts. Bite! Mosquito swarms in the Arctic have driven people temporarily insane. Nine thousand bites per minute will cause a person to lose half of his or her blood supply within two hours.

Mosquitoes have killed with malaria and yellow fever more people than died in all of humankind's wars. In the 1940s and early 1950s three hundred million cases of malaria per year were reported with three million deaths annually. Four million cases were reported in the United States every year until the mid 1930s. Today some mosquitoes in Minnesota carry western equine encephalitis. Thirty cases have been reported in the state since 1975. Twelve Minnesota species carry the dog heartworm parasite.

What to do? You could always try the bear grease and skunk oil mixture that the French Canadian voyageurs slathered on their bodies in the eighteenth century. It's said that it not only kept the mosquitoes away but it also repelled Pierre, Jacques, and Jean Baptiste! Some swear by vitamin B1 pills (thiamine chloride), others by repellants containing DEET (diethyl toluamide), and still others are faithful to Avon's "Skin So Soft." Long pants, long sleeves, a good head-net, and a positive attitude are other weapons that should be in every camper's arsenal.

Bug zappers are a ludicrous testimony to the American dream of making a fortune with a stupid idea. Zappers only serve to attract mosquitoes to the general area, where they can then home in on your much more pleasant carbon dioxide scent. Besides that, bug zappers kill thousands of beneficial insects.

Well then, let's drain the marsh. You'll get rid of the harmless homebody Cattail Mosquito, for sure, and in the process create temporary pools that are the breeding grounds of *Aedes vexans,* a voracious human-biter, and possibly *Culex tarsalis,* the encephalitis carrier. I'd rather suffer a mosquito bite or two than mess up an ecosystem or two.

 **Sparky says:** Let a canoe country mosquito bite you—it won't kill you. Now watch closely. A hand lens may come in handy. Watch it pierce the skin. When do you first feel it? Can you see the blood moving through the mosquito? How well does she fly when she's full? How long before the bite starts itching? Now that wasn't so bad, was it?

Blackfly

Simulium species

Pinhead-size black biting fly

Late May and June is blackfly season

Blackflies will find you. They emerge early and are often abundant by mid-May. But as every canoeist knows, some years the blackfly invasion is worse than in others.

Blackflies use four slashing teeth to cut a shallow well in your skin. Blood fills the gash, and the female blackfly laps it up. She, like the female mosquito, needs a blood meal to provide protein for the developing eggs. Her saliva contains a poison that affects nerve centers and an anticoagulant that inhibits clotting. Welts and intense itching result. As few as twenty to thirty bites may cause a fever of 104° F in some people. Swarming by the millions in the high Arctic, blackflies have been known to kill Barren Ground Caribou.

Approximately three hundred species of blackfly inhabit North America, and about three thousand species are known in the world. (That's one family reunion I wouldn't care to attend!) Also known as humpback gnats, buffalo gnats, and boxers, the blackfly is a pinhead-size true fly with a small down-pointed head that gives it the hunchback look. Some species have white segments on their legs that look like tiny boxing gloves.

The black and blue connection

One day in late May I was in charge of a crew working in deep boreal woods. A fog of blackflies filled the morning air. We all wore long pants tucked into our socks, long-sleeved shirts, hats, gloves, and bandanas on our faces. But I had neglected to tell the crew one thing: blackflies love the color blue. My khaki pants were nearly fly-free, but their blue jeans were nearly black from the living mass of blackflies swarming over them. The key to this blue attraction seems to be in their vision, which is sensitive to light at the low end of the spectrum.

Body heat also attracts the little buggers. Like a heat-seeking missile, they zero in on their target. Their state-of-the-art equipment includes temperature-sensitive cells in the tips of their antennae.

Black and blue II

Blackflies are attracted to another type of "blue"—blueberry flowers. Research by the Ontario Ministry of Natural Resources indicates that blackflies are a major pollinator of the tiny white flowers that later become our beloved, fat, juicy canoe country blueberries.

Sparky says: Remember and give thanks for the blackfly the next time you pick a bunch of blueberries. Without blackflies and their pollinating activities, we would never enjoy the summer's blueberry bounty.

Fast and furious

Fast flowing streams are a home to the larval and pupal stages of the blackfly, providing them with ample oxygen and food. Attached to stream-bottom rocks by suckers, they filter food from the water rushing past them. Mouth-spun silk acts as a safety line if one becomes dislodged and enables it to "reel" itself back to the anchor rock.

The presence of blackflies indicates excellent water quality, since nearly any pollution will kill off the larva. Dry summers and autumns, leading to low stream flow and hence lack of blackfly larva habitat, result in lower populations come next spring.

Deerfly
Horsefly
Family Tabanidae

Deerfly

Length 3/8 inch

Metallic green, gold, or copper eyes and clear wings marked with smoky bands

Horsefly

Nearly 3/4 inch long, with clear wings

The general shape of deerflies, as seen from above, is triangular. Three-eighths inch long with metallic green, gold, or copper eyes, the deerfly has clear wings, marked with smoky bands, that span a half inch at rest. They are beautiful flies in every way but behavior. These are the culprits that incessantly buzz your head while you are defenseless, burdened by an eighty-pound Alumacraft on your shoulders and a #3 Duluth Pack on your back. I especially hate it when they get stuck in my hair. Their ancestral host, the White-tailed Deer, is similarly attacked about the head.

Horseflies, in contrast, are ankle biters. They may be more appropriately called the "mooseflies" as the Moose was likely their ancestral host long before horses set hoof on the continent. They attack all large mammals on their lower extremities. Remember, that it is only the female deerfly and horsefly that bite. Males feed on flowers and are rarely seen. Horseflies are close relatives of deerflies. Both are in the family Tabanidae. Horseflies are much larger, though, often reaching one inch in length, with a stout body and clear wings.

Patriotic flies of freedom

That Americans celebrate the Fourth of July as their Independence Day is the direct result of the horsefly. Folklore has it that on that one July 4, 1776, our forefathers cut short any further debate and discussion on the Declaration of Independence due to the siege of fiercely-biting horseflies in Philadelphia. Desiring to quickly adjourn, delegates decided to sign the document that very day.

Sparky says: Test the hypothesis that horseflies prefer wet skin over dry. Wet one leg and leave the other dry. Which leg interests the horsefly more? Do deerflies show any preference for wet skin?

T he lasagna tasted fantastic, but I couldn't take my eyes off the huge black-and-white bee that was snatching mosquitoes off the window screen. It would hover behind the fly or mosquito, then literally pounce on it and fly off with its treasure, only to return a short while later. I called them "black-and-white bees" then, but have since learned their true identity—the Bald-faced Hornet.

Both the Bald-faced Hornet and the Yellowjacket are members of the family Vespidae, the social wasps. Large-bodied, narrow-waisted, nearly hairless wasps, they make paper nests. At rest they hold their wings back, parallel to the sides of the body. Neither is aggressive away from the nest.

The inch-long Baldface is mostly black, marked with white on the face, thorax, and tip of the abdomen. Walk in the woods and you may see the paper nests of Bald-faced Hornets attached pendulously to the outer branches of deciduous trees. Winnie the Pooh can try all he likes to get honey from one, but he'll never succeed. Only honeybees make honey and they nest only in hollow trees or human-provided apiaries.

Yellowjackets, by contrast, are 1/2 to 3/4 of an inch long and strongly marked with black and yellow. They construct their paper nests underground, often in abandoned mouse nests. The small entrance hole lies amongst the exposed roots of a standing tree. Yellowjackets cause humans more trouble (i.e., sting more) because their nests are concealed and we regularly walk past and over them. Do you blame them for defending their home?

Hornets at home (Yellowjackets too)

Fertilized Bald-faced Hornet and Yellowjacket queens are scattered about the January woods, frozen and hibernating beneath the snows, like seeds waiting for the warmth of spring. From each queen will sprout a new colony, a thriving city of up to fifteen thousand, centered at the paper nest.

The queen's first spring chore is to build several paper cells

Bald-faced Hornet
Vespula maculata

Yellowjacket
Vespula maculifrons

Bald-faced hornet

May reach 1 inch long

Black and white striped

Make paper nests in tree branches

Yellowjacket

1/2 inch to 3/4 inch long

Marked with black and yellow

Nest underground

189

and deposit a fertilized egg in each. Hanging upside down, the plump gray grublike wasp larvae are eating machines. A vegetarian herself, surviving on nectar and tree sap, the queen must go hunting for insects to feed them. The mosquitoes or flies she catches are not eaten but taken back to the nest, chewed up, and formed into "meatloaf," which are then offered to the larvae. The reward to the queen is a drop of very sweet liquid secreted by the larvae. All her time and energy goes into the care of her developing offspring.

Soon, though, the larvae secrete a cocoon and metamorphose into female workers. They take over the feeding of the young and building of the nest, freeing the queen to devote herself solely to egg laying. The colony grows, as does the nest. Tearing up old outer shells, gathering wood pulp, making paper, fixing damaged cells, constructing new cells, cleaning the nest, gathering food, and feeding the larvae and queen make up a roster of rotating jobs the female workers tackle in apparent harmony, as if some kind of "crew boss" were on duty at all times. Princeton professor Kenneth Davis likens the nest community to a living organism, with no wasp acting independently but all working together to keep it alive.

Drones, male wasps hatched from unfertilized eggs, are as highly specialized as the queen, performing no role within the community except to impregnate young queens after they pupate in late July and August. The old queen by now is running on empty. No more eggs. No more larvae. The social order of "the organism" breaks down. By late summer the workers, drones, and old queen have died, their mission completed, leaving only the young fertilized queens to carry on.

Pioneers of paper production

The old weathered wood of BWCAW campsite latrines makes excellent pulp for the construction of wasp nests. Bald-faced Hornets land on the latrine, and with their mandibles they rasp off a strip of wood as they back down the sides. They ball up the pulp and carry it back to the nest. Mixed with saliva, the wood fibers are plastered around the interior hexagonal cells, encasing them. This is papermaking in its most basic form.

The hexagonal cells themselves are also made of wood pulp and saliva. The six-sided shape is the most efficient design for creating maximum space using a minimum of materials. Adjoining cells share walls. The wasp nest is a true architectural marvel, not only for its economy of materials but also for its use of ventilation ducts and insulating air pockets, all incorporated into a balance of structural stresses.

In 1710 René Antoine Ferchault de Réaumur, a French naturalist and physicist, studied the nests of American wasps and determined that it was possible to make paper from wood—a radical discovery, since all paper up to that time was made of rags and linen. No one acted upon Ferchault's claim until some fifty years later, when Dr. Jacob Christian Schaffer, a Bavarian clergyman, began to create many types of paper from a variety of plant material. Today pulp logging for papermaking is a billion-dollar industry in both Minnesota and Ontario.

Sparky says: Check out your BWCAW campsite's government latrine. Look for the tiny vertical scarring on the sides that attests to the pulp gathering efforts of local papermaking wasps. If your campsite has been equipped with the United States Forest Service's new deluxe fiberglass model, you're out of luck and so are the wasps.

Fisher Spider

Dolomedes species

Very large spider

Long legs span 5 inches

Often found on lichen-covered rocks near lake shores

Can run on top of water and dive below

The fisher spider is a large, water-loving arachnid. Along with other spiders, mites, scorpions, and ticks, it's one of the air-breathing, eight-legged, antennaless invertebrates of the class Arachnida.

By far the largest spider in Minnesota, the fisher spider may have a body as large as $1\frac{1}{2}$ inches, with legs spanning 5 inches. Fisher spiders are so sexually dimorphic that several scientists have mistakenly described them as separate species. Males are often only half as large as females, with considerably smaller abdomens and a conspicuous yellow band around the forebody. Females are marked with two yellowish bands that run the length of the body and three to six pairs of light spots.

Lichen-encrusted boulders, just above waterline, provide the preferred watching perch for this fishing spider.

Diving Dolomedes, the amazing fish-eating spider

A case has been documented of a $3/4$-inch fisher spider catching a $3\frac{1}{2}$-inch fish that weighed four times more than the spider. It dragged the fish from the water and ate it. That's like a human diving into the ocean, grabbing a 600-pound Steller Sea Lion, pulling it to shore, and eating it raw. The minnow was a seven-course gastronomic feast for that little spider. The fisher spider's everyday bread-and-butter, though, consists of small aquatic insects, tadpoles, tiny land critters, and fish fry, which they pursue and capture without the aid of a web.

Fisher spiders possess the amazing ability to dive underwater to pursue prey or to escape a predator. Bubbles trapped in their highly water-repellant body hairs provide enough breathing air for some species to stay under thirty minutes or more. These same water-repellant hairs allow fisher spiders to run across the surface of the water. They can also swim across the surface, using a paired-leg breast stroke.

Nursery webs

Though Dolomedes spiders prefer to pursue-and-pierce rather than wait-and-web their victims, they still do have a use for their silk-producing spinneret organs. Some species use them to create homey little nursery webs for the protection of newly hatched young.

The female lays about three hundred eggs in a sac nearly as large as her abdomen; she awkwardly carries it under her, firmly grasped in her jaws. Just before hatching time, she attaches the egg sac to leaves, tears it open, and spins a nursery web around it. At first, as the young are freed they begin to eat one another. Mom guards nearby but does not feed them. They leave the nursery web as they grow and can better fend for themselves.

Sparky says: With headlamp on, slip your canoe into the dark night waters. Shine your light along the rocky shore edge, watching for the green glow of the fisher spider's eight eyes.

Spider Woman

When a woman named Arachne challenged the Greek goddess Athena to a weaving contest, Athena punished her by turning her into a spider. Arachne is the inspiration for the naming of the class to which the spiders belong—Arachnida.

Wood Tick

Dermacentor variabilis

Deer Tick

Ixodes dammini

Wood Tick

Tiny oval-body ³/16 inch long

Eight short legs

Dark back marked with two light stripes or horseshoe mark

Deer Tick

Pin-head size

Orange-brown body with a dark spot near head

Deer tick / Wood Tick
Actual size

W hen I first started working in the canoe country, with great pride we told nervous visitors that they'd encounter no poisonous snakes, no Poison Ivy, and hardly a Wood Tick. That was 1981. Now in the early 1990s, Wood Ticks seem to be abundant, in grassy, brushy areas such as dried beaver ponds. Why the explosion? Could it be a series of dry summers? I don't know, but Superior National Forest wildlife biologist Ed Lindquist agrees that Wood Ticks are apparently more common today than in the early 1980s. No systematic survey has quantified the tick increase, though.

Wood Ticks differ from Deer Ticks in being much larger in every stage of life. The oval bodies of adult Wood Ticks are ³/16 inch long. You can tell the male and females apart by the markings on their dark backs. The male Wood Tick sports suspenders (two light stripes down the back), while the female Wood Tick wears an apron (horseshoe-shaped mark).

The adult Deer Tick is barely the size of a pinhead. If you get close and squint you may be able to see that it has an orange-brown body with a dark spot near the head.

The "tick-ing" clock

Wood Ticks have a one-year life cycle. In the spring adults climb up to knee-height in grass (they do not drop out of trees) and hold on tight with six of their eight legs. They hold out the front hooked legs, waiting to grasp the first large mammal to saunter by. Moose, deer, fox, wolf, Rover, and you and I fit the bill. On the host, the tick gathers up a square meal of blood. If she has her way, she'll gorge till she's the size of a Junebug, then release her jaws and fall off. Just like the mosquito, she needs this blood meal for the development of her eggs.

Three thousand to six thousand eggs are laid. They hatch in one to two weeks. The six-legged microscopic larva finds a tiny mammal to play vampire with, then drops off, and metamorphoses into an eight-legged nymph. Blood. Drop. Adult. Survive the winter. Mate. The cycle starts anew.

Lyme disease

Deer Ticks have always been here but they never received any attention until their cousins in Old Lyme, Connecticut, started passing a nasty disease on to humans. Lyme disease may result in a red "bullseye" rash, flulike symptoms, aches and pains, arthritis, withdrawal, lethargy, Bell's palsy, and, in rare cases, death. However, Lyme disease is not common, and suffering symptoms beyond the rash is extremely rare. The first reported case of the disease in the Boundary Waters–Quetico area occurred in the late 1980s.

Corkscrew-shaped bacteria called spirochetes that are transmitted to humans by infected Deer Tick bites cause Lyme disease. Here's how it works. The Deer Tick's life cycle is essentially the same as the Wood Tick's, but takes two full years. Adult ticks feed mainly on White-tailed Deer. They gorge, fall off, lay eggs, and die. The eggs hatch. The resulting larva may pick up the Lyme disease spirochete from an infected host. The main host for immature Deer Ticks is the White-footed Mouse, which is completely unaffected by the disease. The infected nymphs turn into adults and wait patiently for the next large mammal host. If they can bite and stay attached for several hours, they will transmit the Lyme disease spirochete.

A European parasitic wasp may help control the tick, and hence the disease. Burton Engley, a 76-year-old Martha's Vineyard resident who has suffered two mild cases of Lyme disease, tracked down a report on a 1926 experiment in which Chalcid Wasps were released to kill Wood Ticks. The Chalcid Wasps lay their eggs inside the bodies of ticks. When the wasp eggs hatch they eat their way out, killing the tick host. A 1985 study confirmed that the Chalcid Wasps were still infesting and killing 27 percent of the area's ticks. It sounds great, but introducing one foreign species to control another has a poor track record and makes some people more than a little nervous.

Diagnosis of Lyme disease is still tricky, as blood tests are often inconclusive. The people at 3M in St. Paul are now researching a urine-antigen test. Researcher Dr. Russell Johnson at the University of Minnesota has discovered antigens in the urine of infected mice and humans, but the consistency of this presence has yet to be proven. With so much research being done on Lyme disease, those who live or travel in tick country can hope that just over the next horizon are a vaccine, a reliable diagnostic test, and effective antibiotics.

Remember that, for now, Lyme disease is a rarity in the canoe country, but long pants, long sleeves, DEET repellant, and careful tick checks can add to your peace of mind.

 Sparky says: Try your hand at sexing Wood Ticks. Remember the male wears suspenders and the female an apron. These distinctive light markings on the dark body are especially easy to see with a hand lens. Do you find more males or females? Why?

Yuck! Slimy, bloodsucking pests that ruin all chances of enjoying a swim. What could possibly be interesting about leeches? They don't even have eyes!

Many species of leeches undulate their way through canoe country waters. The large American Medicinal Leech is one of the most common. Its 3 to 6 inch long olive body is 1/2 inch wide, with orange central dots and orange undersides. Amazingly, fish find this leech repulsive. The Ribbon Leech, also abundant, is the most common bait leech, relished by many species of fish. It is 3 to 4 inches long, gray-brown blotched with black. Leeches are abundant in midsummer and are gorged on by Northern Pike, Walleyes, and Smallmouth Bass.

The leech you find attached to your leg after a swim needs blood for its survival. A sucker on each end of its body secures itself to you. Within the "head" sucker, three razor-sharp serrated jaw plates slice through the skin in a jigsaw motion. As the blood begins to flow, the leech injects the anticoagulant hirudin into the wound. The blood is dehydrated and stored in intestinal sacs. Symbiotic bacteria prevent decay, allowing some leech species to survive a full year on one blood meal. Once a leech is fully gorged, it will fall off. You can speed the process with a good yank or a liberal application of salt.

Not all leeches suck mammalian blood. Horse Leeches, *Haemopis* species, feed on aquatic worms, insect larvae, carrion, and organic mud. Three or four families of smaller leeches have no jaw plates, but instead have retractable proboscises to pierce and suck snails and clams. In South America and southeast Asia there even dwell land leeches.

Leeches are hermaphroditic. They have both ovaries and testes and can fertilize themselves. But to avoid the scourge of inbreeding they find sex partners.

The doctor and the leech

The leech has been used as a medicinal tool for centuries. The only cure for "bad blood," the supposed cause of many an

Leech
Family Hirudinea

Average length 2 to 4 inches

Olive, brown, black, or gray

Soft-bodied, and flat

Undulate their way through lakes, ponds, and streams

May attach to human skin in order to suck blood

Gross!

In 1925 the polluted Illinois River produced 24,336 leeches per square yard. That's the equivalent of a ton of leeches per acre.

Of all the nerve

The giant nerve ganglia of leeches provide physiologists with a wonderful tool for studies of the nervous system.

197

Varicose veins

Hirudin, the anticoagulant gathered from the leeches' buccal glands, is a basic ingredient in ointments to treat varicose veins.

Acid rain and leeches

Leeches are one of the most acid-sensitive freshwater organisms around. A study by Berry Bendell and Don McNicol on forty small lakes near Sudbury, Ontario, came to this conclusion: as the pH level of the water drops from greater than 5.5 to less than 4.9 (indicating increased acidity) the leech population dies out.

Leeches are a basic component of the food chain, a necessary fish food, and also crucial recyclers. By breaking down dead organic matter, they release nutrients, making them available for plants and other aquatic organisms.

ailment, was a good bloodletting. Both physician and patient often preferred leeches to the ugly alternative of a rusty lance. Six leeches on each temple was the remedy for a common cold.

European peasants gathered leeches on their own legs and sold them. By 1848, overharvest in Russia led to leech laws and a closed season from May to July. The upshot in the rest of Europe was a sixteenfold increase in leech prices. Only the wealthy could then afford the leech treatment.

Today the wealthy and the well-insured still benefit from leech technology in medicine. America and France use leeches as an adjunct to microsurgery. For example, when connecting a severed finger, surgeons use leeches to suck accumulated blood from unconnected microveins. The use of a slimy little leech during delicate, highly sterile modern microsurgery presents a fascinating incongruity.

America's royalty, the actors and actresses of Hollywood, take advantage of the leeches' bloodsucking ability by attaching them near black eyes that need to disappear overnight. They are fast and efficient, and leave only a tiny puncture wound as they suck out the excess blood.

Researchers at the Mayo Clinic in Rochester, Minnesota, have discovered that hirudin, the leech's anticoagulent, may actually penetrate clots in the heart's arteries, preventing them from growing, and thereby possibly preventing heart attacks. The research continues.

Sparky says: Snorkel the shallows around your campsite. Check out aquatic critters in their natural habitat, including leeches, bass, dragonfly nymphs, and crayfish. Explore the weedy patches, boulder fields, and lake bottom. A quality mask and snorkel are essential, but fins are optional.

Walter Momat of Thunder Bay's Lakehead University believes that crayfish are many lakes' dominant organism. The volume of crayfish is directly proportional to the volume of fish. The simplified formula is 10 pounds crayfish = 1 pound fish. A favorite of Smallmouth Bass, crayfish are also eaten by Northern Pike and Walleye. Mink, River Otter, and Great Blue Herons also eat them.

The crayfish excavates a little cave under a protective rock or log and backs into it, leaving its eyes and proportionately massive pincers exposed. By night the crayfish moves out to forage. It seeks out protein-rich carrion, fish eggs, and insect larvae, but the bulk of its diet is periphyton, or "scum flora," a complex green or brown film of organic detritus, bacteria, and algae found on lake rocks.

Several species of crayfish inhabit the canoe country. Few exceed 5 inches in length. The best time to see them is at night amongst the rocks of the lake's bottom. A flashlight will aid in the search.

Crayfish
Orconectes species

Less than 5 inches in length

Resembles a tiny orange lobster complete with pincer claws and a fan tail

Poor person's lobster

Many people know that "crawfish" form an important part of Cajun cuisine. Louisiana alone consumes 100 million pounds per year. I was surprised to discover, however, that Duluth, Minnesota, has a crayfish exporting company, Freshwater Crayfish Ltd. I talked with owner Ken Kallagher about the operation. Since the upper midwest has not gotten beyond the "fish stick stage" of seafood appreciation, the company exports all its harvest to Sweden, where crayfish is a delicacy prepared just like lobster.

Sweden's, and most of Europe's, crayfish population has been decimated in recent years by the fungus *Aphanomyces astaci,* which causes a deadly crayfish plague. After the loss of Sweden's crayfish industry, Turkey became the major European supplier. But the crayfish plague spread to Turkey and has totally put them out of business.

Duluth's Freshwater Crayfish Ltd. buys baited and trapped crayfish in northern Wisconsin from June through October. They buy only the large species, the Rusty Crayfish and the Fantail Crayfish (which is common throughout the Boundary Waters and Quetico). Sweden prefers the Rusty Crayfish because it turns bright red when boiled, has sweet meat, and has large claws, which means more meat. Freshwater Crayfish Ltd. shipped 11 metric tons to Sweden in 1990.

 Sparky says: Put your headlamp on, get out of the tent, and walk down to the dark water's edge. Night is when crayfish are active. Search the rocky bottom for activity. Sit and observe. Or if you're brave, catch a few, fire up the stove, and immediately drop the crayfish into the heavily salted, boiling water. Boil until deep red, or for one minute. Remove. Twist the tails free and break the shell away from the sweet meat. Dip in melted butter. Mmm—North Wood's lobster!

Appendix

North Woods Primer

Ice Sculpture. Twelve thousand years ago an immense monster gouged, scraped, raked, crushed, and generally mauled the canoe country.

For the fourth time during the 2-million-year Quaternary period, the monster came from the north, slowly and steadily crushing everything in its path.

The monster, a massive glacier 2 miles thick, ground its way south on a base of accumulated granite debris, exerting unstoppable, landscape-altering force.

The glacier had formed thousands of years earlier near the polar cap. As the climate cooled, more snow fell each year than could melt. Sedimentary layers of snow built up, metamorphosing the bottom layers to ice. When the layers reached about 300 feet, gravity caused the mass to bulge and deform under its own weight, starting a southward flow.

About 10,500 years ago a warming climate caused the fourth glacier to slowly begin melting. The retreating glacier revealed a barren earth: white, clean granite inundated by rivers of debris-laden meltwater and dotted with huge blocks of glacier ice. No trees, shrubs, or moss grew. No birds flew in the air. And no mammals prowled the fringe of the ice sheet.

But soon lichens came to colonize the rock. Wind delivered the spores of mosses, and traveling birds dropped the seeds of spruces. Cold air flowing over the glacier created a tundra fringe at its margin. Mammals migrated north.

Imagine the young canoe country grazed by mammoths, mastodons, Ground Sloths, llamas, camels, Giant Beavers, Giant Short-faced Bears, Saber-toothed Cats, Woodland Musk-oxen, and Giant Armadillos. Unfortunately, mysterious mass extinctions eliminated these creatures from North America's tundra lands forever.

The climate continued to moderate, allowing more southerly plant and animal species to invade. The glaciers retreated north, receding at a rate of 1 inch to 10 feet per day. Meanwhile, in the canoe country, meltwater filled gouged granite depressions to form the thousands of lakes, ponds, and rivers that make it a canoeist's paradise. Today, a portion of this unique watery landscape is preserved as the Boundary Waters Canoe Area Wilderness and the Quetico Provincial Park.

The spruce-moose forest today

The great North Woods (of which the Boundary Waters and Quetico is the southern edge), one of the most recent glacially uncovered places on the face of the globe, sits atop some of the world's oldest rocks. The granitic Canadian Shield, which underlies 1,864,000 square miles of Canada and the western Great Lakes in a broad arc around Hudson Bay from the Atlantic to the Arctic Ocean, is one of the few surviving chunks of the Earth's original crust. Lavas which extruded from beneath the surface 2.7 billion years ago cooled slowly to form a gigantic granite batholith. In spots, movement along faults shoved this mass skyward, creating a massive mountain range in what is today northeastern Minnesota and northwest Ontario. Successive glaciers ground the peaks down to the low hills we see today, mere roots of the once mighty mountains.

The northern forest, which covers much of the Canadian Shield, was named for Boreas, the Greek god of the north wind. The boreal forest, a land of spruce, fir, and birch, is home to Pine Marten, Moose, and mergansers. Characterized by hot wet summers and long cold snowy winters, the Boundary Waters and Quetico receive an average of 27 inches of precipitation annually, 60 percent coming in the form of rain. Snow makes up the rest, 60 inches annually, and blankets the ground for five months, from mid-November to mid-April. Lagging a half month behind the snow, large deep lakes usually freeze over by early December and break up in early May. Temperatures can range from 100° F to -50° F. January, the coldest month, averages 12° F. July, the warmest, averages 62° F. Swimming in Boundary Waters and Quetico lakes becomes comfortable (i.e., bearable) by the fourth of July.

Nicknamed the spruce-moose forest, the boreal forest is, as one author put it, the seam between two worlds. It divides the northern tundra from the deciduous forests of the south, not only on this continent but in Europe and Asia as well. The globe wears the boreal forest around its head like a spruce-studded crown. You could wake up on five consecutive mornings in northern Alberta, Finland, Ukraine, northern China, and the Boundary Waters or Quetico and never realize you'd left the canoe country. Many boreal species are circumpolar and can be found in northern latitudes around the world. Moose (called elk in Europe), Lynx, Three-toed Woodpeckers, Great Gray Owls, and Caribou Lichen are all examples of circumpolar species. If you look closely though, you'll notice that Eurasian boreal forests are primarily composed of similar but slightly different species of spruces, firs, jays, woodpeckers, weasels, mosses, and lichens.

Only an average of 10 inches of soil has built up on top of the Boundary Waters and Quetico bedrock granite since the last glacier melted 10,000 years ago, but this thin skin of soil can support and feed trees well over 100 feet tall. Most large tree species have spreading roots that lace together with the roots of other trees to form an interwoven mat. When high winds topple such a tree, it lifts up the root mat and soil, peeling it cleanly off the underlying bedrock, forming what some call a witch's cradle.

Fire—friend or foe?

If all fire, wind, logging, and disease were eliminated from the canoe country, the upland forests would climax as a White Spruce–Balsam Fir–Northern White Cedar mosaic, with Black Spruce and Tamarack dominating the bogs. As climax forests they would regenerate themselves indefinitely. But natural disasters are a part of the boreal ecosystem. They are creative rather than destructive forces, renewing the land.

Smokey Bear has warned us for years about the dangers of fire, but all fire is not bad. A natural rejuvenating force, fire has been part of the natural landscape for millennia. Ecologist

Bud Heinselman did extensive tree coring in the BWCAW to determine the area's natural fire rotation. By counting tree rings and studying fire scars, Heinselman determined that approximately every 100 years during presettlement times, the bulk of BWCAW lands burned over at least once. Drought helped make the years 1681, 1692, 1727, 1755-59, 1801, 1824, 1863-64, 1875, and 1894 the canoe country's biggest fire years. Climax spruce-fir-cedar forests were the exception and not the norm before the coming of Smokey and this century's history of fire suppression.

The spruce-moose forest is, and always has been, fire dependent. Let's say our climax forest burns in a lightning-ignited fire. Within two weeks new green plants will push through the blackened earth. I witnessed this phenomenon 19 days after the 440-acre Bearskin Fire of May 1988. Clintonia, Fringed Bindweed, and Large-leaved Aster thrived in the scorched soils now rich in newly released phosphorous, magnesium, and calcium. By the end of the summer Quaking Aspen roots had sent up thousands of 4-foot-tall suckers. Aspen and Jack Pine are pioneer species that flourish on sunny, burned-over lands. The serotinous (resin-closed) cones of Jack Pine open in the intense heat of a wildfire, releasing millions of seeds onto the competition-free, nutrient-rich soil. After Minnesota's 1971 Little Sioux fire, researchers discovered an amazing average of 39,910 Jack Pine seedlings per acre.

Twenty years after a fire, aspen and birch dominate the canopy. Shade-tolerant White Spruce and Balsam Fir sit quietly below, waiting for the short-lived aspens and birches to topple. A climax forest is the result.

Though protected from ground fires by thick bark, the massive Red and White Pines are nonetheless fragile creatures. Their start depends on a good seed supply, bare soil, sunlight, and ground fires to clean out faster-growing competitors. But once established, pine stands can reach tremendous age, as demonstrated by the Red Pines on Seagull Lake's Three Mile Island. Originating after a 1595 fire, these oldest trees in the

BWCAW will celebrate their four hundredth birthday in 1995. Red and White Pines are not climax species as many suspect, for when they topple from old age, the spruces and firs again reclaim the canopy.

Today Smokey Bear is changing his tune. He's realizing the value of a good fire to the well-being of the forest. Besides cleaning up the forest-floor debris and releasing nutrients to the soil, fire helps Moose, Beavers, Snowshoe Hare, and White-tailed Deer by creating succulent new browse. Concentrations of these herbivores attract hungry Timber Wolves. Bears and Pine Marten show up to gorge on abundant blueberries and raspberries. Seeds galore attract small rodents, which in turn attract Broad-winged Hawks and American Kestrels. Three-toed and Black-backed Woodpeckers take advantage of the new abundance of wood-boring beetles. With better under-standing of this ecosystem, the United States Forest Service is now allowing small nonthreatening fires to burn their course within the BWCAW. It's a start. Soon perhaps, fire will again become an integral part of the Boundary Waters' and Quetico's natural history.

Table of Measurements

U.S. Unit	Metric Equivalent	
1 inch	2.54	centimeters
1 foot	.31	meters
1 yard	.91	meters
1 mile	1.61	kilometers
1 acre	.405	hectares
1 ounce	28.35	grams
1 pound	453.6	grams
1 pound	.45	kilograms

Metric Unit	U.S. Equivalent	
1 centimeter	.39	inches
1 meter	3.28	feet
1 meter	1.09	yards
1 kilometer	.62	miles
1 hectare	2.45	acres
1 gram	.035	ounces
1 kilogram	2.2	pounds

Keeping Track of Nature

Memories fade quickly, and soon one trip blends into another. Written and photographic records have helped me keep fresh my encounters with wildlife. You may wish to use the worksheets on the following pages to keep track of the details—time and place, weather, and activities—of the animals you observe. You are welcome to photocopy the worksheets for yourself and for your group.

Observation Checklist

Make observing wildlife a game, make it a challenge, or just do it for fun. Keep track of each species that you see or hear while on your trip. Check each off on the Observation Checklist, then make a few notes about the sightings. After you return home your notes will help you relive the experience again and again. You'll remember that Red Squirrel much better if you not only check it off but also note that it woke you up the first day of your journey by dropping cones on your tent.

You may want to keep a checklist for each trip, plus a master list that records your lifetime of wildlife observation.

Phenology Journal

Spring was just mud puddles in my driveway until I discovered the hobby and science of phenology. As I learned to pay attention to the predictable rhythm of the seasons I began to look forward to spring, anxiously await summer, fling myself into fall, and, yes, even embrace winter. The subtle became the obvious. Seasons became a complex conglomerate of specific natural events.

Phenology is the science that studies the timing of natural events from year to year and place to place, and their relationship to weather and climate. In my phenology calendar I record the first Common Loon back on the lake in spring, the blooming of the first Twinflower, the Walleye spawn, the Sugar Maple sap run, the turning color of the Tamarack needles, the first callings of the American Toad, ice-up, ice-out, birth of the Woodchuck kits, Luna Moths fluttering at the porch light, the first Ruby-throated Hummingbird at the feeder, the first ripe blueberry, the last Black Bear still foraging before hibernation, and a zillion other natural seasonal occurances.

I list these events in an 8 1/2-by-11-inch hard-covered blank sketchbook. I divide each page into three horizontal sections. One page covers one date over three different years. For example, I have one page in my current phenology journal for May 27. Natural events and weather data for May 27, 1992, are in the top section; below that is the 1993 section; and the bottom is reserved for the same date in 1994.

You'll find it fascinating to compare your nature observations from year to year. The seasons seem to progress at varied rates each separate year. But over the years you'll discover that many events actually occur with clockwork regularity.

Start your phenology journal today. You can start small by jotting observations on a wall calendar. Use the phenology calendar that follows to write down nature observations from your canoe trips.

Daily Timetable

Animals establish a predictable daily routine that includes eating, sleeping, working, and playing, just as humans do.

You'll rarely see Whitetailed Deer feeding during the heat of midday. As automobile drivers know, dawn and dusk are peak times for deer activity. Anglers are aware that some hours of the day are typically better than others for fishing. And everyone knows that mosquitos are most ferocious at dusk.

Begin to take notice of the rhythms within the day. Listen for bird songs—when does the White-throated Sparrow sing? If you catch a fish, record the time. Perhaps that information will help you catch another one. Your records will help you become aware of the daily patterns of wildlife.

Observation

Check off the wildlife you see

Mammals

- ❏ Bats
- ❏ Black Bear
- ❏ River Otter
- ❏ Pine Marten
- ❏ Mink
- ❏ Timber Wolf
- ❏ Red Squirrel
- ❏ Northern Flying Squirrel
- ❏ Eastern Chipmunk
- ❏ Least Chipmunk
- ❏ Beaver
- ❏ Red-backed Vole
- ❏ White-tailed Deer
- ❏ Moose

Birds

- ❏ Common Loon
- ❏ Common Merganser
- ❏ Mallard Duck
- ❏ Black Duck
- ❏ Great Blue Heron
- ❏ Spotted Sandpiper
- ❏ Herring Gull
- ❏ Turkey Vulture
- ❏ Bald Eagle
- ❏ Osprey
- ❏ Broad-winged Hawk
- ❏ Barred Owl
- ❏ Spruce Grouse
- ❏ Ruffed Grouse
- ❏ Common Nighthawk
- ❏ Belted Kingfisher
- ❏ Black-backed Woodpecker
- ❏ Three-toed Woodpecker
- ❏ Pileated Woodpecker
- ❏ Gray Jay
- ❏ Common Raven
- ❏ Black-capped Chickadee
- ❏ Boreal Chickadee
- ❏ Winter Wren

Checklist

nd make notes of your observations.

_____

_____

_____

_____

_____

_____

_____

_____

_____

_____

_____

_____

_____

_____

_____

_____

❑ Swainson's Thrush
❑ Cedar Waxwing
❑ Red-eyed Vireo
❑ Warblers
❑ White-throated Sparrow

Fish
❑ Walleye
❑ Northern Pike
❑ Lake Trout
❑ Smallmouth Bass

Reptiles and Amphibians
❑ Eastern Garter Snake
❑ Snapping Turtle
❑ Western Painted Turtle
❑ Wood Frog
❑ Green Frog
❑ American Toad

Insects and Other Invertebrates
❑ Mayfly
❑ White-tailed Dragonfly
❑ Black-winged Damselfly
❑ Spittlebug (Froghopper)
❑ Water Strider
❑ Sawyer Beetle
❑ Whirligig Beetle
❑ Firefly
❑ Luna Moth
❑ Tiger Swallowtail
❑ Mosquito
❑ Blackfly
❑ Deerfly
❑ Horsefly
❑ Bald-faced Hornet
❑ Yellowjacket
❑ Fisher Spider
❑ Wood Tick
❑ Deer Tick
❑ Leech
❑ Crayfish

Phenology

Record your own observation

April

Week 1: The cry of Herring Gulls can be heard near open rivers and lakes.

Week 2: Black Ducks and Mallards return to open pools.

Week 3: Black Bears emerge from hibernation.

Week 4: Spruce Grouse males perform their elaborate mating dance.

May

Week 1: Wood Frogs and Spring Peepers begin calling from meltwater ponds.

Week 2: Black Bears munch on green grass.

Week 3: Moose cows with twin calves eat Mountain Maple leaves.

Week 4: Watch out! Blackflies hatch.

June

Week 1: Proud Mallard mothers lead flotillas of ducklings.

Week 2: Snapping Turtles lay 20 to 50 Ping-Pong-ball-size eggs in sandy soil.

Week 3: Green Frogs "plunk" and American Toads "trill."

Week 4: "Loonlets," brown and fuzzy, ride on parents' backs.

©1993 Pfeifer-Hamilton Publishers PO Box 3151 Duluth MN 55803 (800) 247-6789

Calendar

of the seasonal patterns.

_____

_____

_____

_____

_____

_____

_____

_____

_____

_____

_____

_____

_____

_____

_____

July

Week 1: Spruce Budworm Moths emerge.

Week 2: Stream shores alive with Black-winged Damselflies.

Week 3: Bird song decreases. The woods are quiet.

Week 4: Red Squirrels harvest Beaked Hazelnuts.

August

Week 1: The sky is falling! No, it's just Red Squirrels dropping cone-laden Jack Pine and spruce branches.

Week 2: Caterpillars of the Luna Moth are creeping around.

Week 3: Mosquitos are gone!

Week 4: Common Nighthawks begin moving south.

September

Week 1: Mushrooms multiply after early fall rains.

Week 2: Snapping Turtle eggs begin to hatch.

Week 3: The hawk migration is at its peak.

Week 4: The gold and red leaves of deciduous trees glow against the deep green of the conifers.

Daily

Record your own notes o

Dawn

_____

_____

_____

_____

_____

_____

Midmorning

_____

_____

_____

_____

_____

Midday

_____

_____

_____

_____

_____

Timetable

he daily rhythms you observe.

_____ **Midafternoon**

_____

_____

_____

_____

_____

_____ **Dusk**

_____

_____

_____

_____

_____

_____ **Late night**

_____

_____

_____

_____

_____

Sources

Alcock, J. 1979. Multiple mating in *alopteryx maculata* and the advantage of non-contact guarding by mates. *Journal of Natural History.* 13:439-440

Aleksiuk, M. 1975. Manitoba's fantastic snake pits. *National Geographic.* 148:715-723.

Baker, M. C., E. Stone, A. E. Miller Baker, R. J. Sheldon, P. Skillicorn, and M. D. Mantyoh. 1988. Evidence against observational learning in storage and recovery of seeds by Black-capped Chickadees. *Auk* 105:492-497.

Barash, David. 1977. Sociobiology of rape in Mallards. *Science* 197:788-789.

Berdell, B. E., and D. K. McNicol. 1991. An assessment of leeches (Hirudinea) as indicators of lake acidification. *Can J. Zool.* 69:130-133.

Bergerud, A.T., and M.W. Gratson. 1988. *Adaptive strategies and population ecology of northern grouse.* University of Minnesota Press, Minneapolis.

Bock, W. J. 1961. Salivary glands in the Gray Jay. *Auk* 78:355-365.

Brodsky, L. M., C. D. Ankney, and D. G. Dennis. 1989. Social experience influences preferences in Black Ducks and Mallards. *Can. J. Zool.* 67:1434-1438.

Burkhardt, D. A. 1984. Probing the secrets of the Walleye's eye. *MN Volunteer* 47(275):36-42.

Dahlberg, B. L., and R. C. Guettinger. 1956. The White-tailed Deer in Wisconsin. *Wisconsin Conservation Department Technical Wildlife Bulletin* Number 14.

Davis, W.J. 1982. Territory Size in *Megaceryle alcyon* along a stream habitat. *Auk* 99:353-362.

Dow, D. D. 1965. The role of saliva in food storage by the Gray Jay. *Auk* 82:139-154.

Ehrlich, P. R., D. S. Dobkin, and D. Wheye. 1988. *The birder's handbook.* Simon and Schuster Inc., New York.

Fellegy, J. 1982. *Classic Minnesota fishing stories.* Waldman House Press, Minneapolis.

Gullion, G. 1984. *Grouse of the North Shore.* Willow Creek Press, Oshkosh, WI.

Hazard, E. B. 1982. *The Mammals of Minnesota.* University of Minnesota Press, Minneapolis.

Johnson, B. 1989. *Familiar amphibians and reptiles of Ontario.* Natural Heritage/Natural History Inc. Toronto.

Kohler, O. 1950. The ability of birds to "count." Bull. *Animal Behaviour* 9:41-45.

Koplin, J. R. 1969. The numerical response of woodpeckers to insect prey in a subalpine forest in Colorado. *Condor* 71:436-438.

Koplin, J. R., and P. H. Baldwin. 1970. Woodpecker predation on an endemic population of Engelmann Spruce Beetles. *Am. Midl. Nat.* 83:510-515.

Krebs, J. R. 1974. Colonial nesting and social feeding as strategies for exploiting food resources in the Great Blue Heron. *Behavior* 51:99-134.

Lawrence, L. de K. 1954. The voluble singer of the tree tops. *Audubon* 56:109-111.

MacArthur, R. H. 1958. Population ecology of some warblers of northeastern coniferous forests. *Ecology* 39:599-619.

McIntyre, J. W. 1988. *The Common Loon:Spirit of northern lakes,* University of Minnesota Press, Minneapolis.

Merrit, J. F. 1981. *Clethrionomys gapperi.* Mann. *Species* 146:1-9.

Mills, A. M. 1986. The influence of moonlight on the behavior of goatsuckers (Caprimulgidae). *Auk* 103:370-378.

Mountjoy, D. J., and R. J. Robertson. 1988. Why are waxwings "waxy"? Delayed plumage maturation in the Cedar Waxwing. *Auk* 105:61-69.

Munro, J. A. 1945. Observations of the loon in the caribou Parkland, British Columbia. *Auk* 62:38-49.

Nagy, J. G., and W. L. Regelin. 1977. Influence of plant volatile oils on food selection by animals. *Int. Congr. Game Biol.* 13:225-30.

Nelson, R.K. 1983. *Make prayers to the ravens.* The University of Chicago Press, Chicago.

Nisbet, I. C. T., W. H. Drury, and J. Baird. 1963. Weight loss during migration. Part 1:Deposition and consumption of fat by the Blackpoll Warbler. *Bird Band* 34:107-138.

Oring, L. W., D. B. Lank, and S. J. Maxson. 1983. Population studies of the polyandrous Spotted Sandpiper. *Auk* 100:272-285.

Owre, O. T., and P. O. Northington. 1961. Indication of the sense of smell in the Turkey Vulture from feeding tests. *Amer. Midland Nat.* 66:200-205.

Robinson, W. L. 1980. Food hen: *The Spruce Grouse on the Yellow Dog Plains.* University of Wisconsin Press, Madison.

Rothstein, S. I. 1971. Observation and experiment in the analysis of interactions between brood parasites and their hosts. *Amer. Nat.* 105:71-74.

Schmid, W. D. 1982. Survival of frogs in low temperatures. *Science* 215:697-698.

Smith, S. A., and R. A. Paselk. 1986. Olfactory sensitivity of the Turkey Vulture to three carrion-associated odorants. *Auk* 103:586-592.

Southern, W. E. 1958. Nesting of the Red-eyed Vireo in the Douglas Lake region, Michigan. *The Jack-Pine Warbler* 36:105-130, 185-207.

Stager, K. E. 1964. The role of olfaction in food location by the Turkey Vulture. Los Angeles County Mus., *Contrib. in Sci.* 81:1-63.

Strom, D. 1986. *Birdwatching with American Women.* Norton, New York.

Swift, R. W. 1948. Deer select most nutritious forages. *Journal of Wildlife Management* 12(1):109-110.

Tinbergen, N. 1953. *The Herring Gull's world.* Collins, London.

Truslow, F. K. 1967. Egg-carrying in the Pileated Woodpecker. *Living Bird* 6:227-235.

Waage, J. K. 1979. Dual function of the damselfly penis:Sperm removal and transfer. *Science* 203:916-918.

Yack, J. E. 1988. Seasonal partitioning of atympanate moths in relation to bat activity. *Can. J. Zool.* 66:753-755.

Other fine books from ℞ Pfeifer-Hamilton Publishers

CampSights
Sam Cook

The third delightful collection of essays and stories from the North Country. Sam offers insights into the subtleties of the natural world that all too often go unnoticed. Sam's first two books, *Up North* and *Quiet Magic*, are also available from Pfeifer-Hamilton Publishers.

Hardcover, 208 pages, $16.95

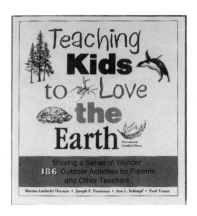

Teaching Kids to Love the Earth
Marina L. Herman, Ann Schimpf, Joseph Passineau, and Paul Treuer

A collection of 186 earth-caring activities designed for use with children of all ages to help them experience and appreciate the Earth. *Teaching Kids to Love the Earth* will enable you and the children you work with to experience a sense of wonder about the world we all share.

Softcover, 192 pages, $14.95

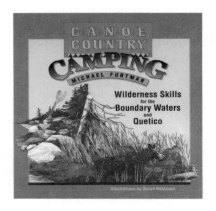

Canoe Country Camping: Wilderness Skills for the Boundary Waters and Quetico
Michael Furtman

Entertaining, up-to-date, and complete, *Canoe Country Camping* has everything you'll need to enjoy a safe and fun-filled camping trip. You will find detailed drawings, helpful charts, and handy checklists. *Canoe Country Camping* is a great gift for seasoned canoeists as well as beginners.

Softcover, 216 pages, $14.95

The North Shore:
A Four-Season Guide to Minnesota's Favorite Destination
Shawn Perich

Explore Minnesota's North Shore with this personal tour-guide. *The North Shore* will help you plan your adventures, from one-day excursions to two-week vacations. Read *The North Shore* as you plan your trip; then take it along to enjoy the milepost-by-milepost descriptions of Lake Superior's scenic splendor.

Softcover, 216 pages, $14.95

Gunflint: Reflections on the Trail
Justine Kerfoot

Justine Kerfoot has lived on Minnesota's remote Gunflint Trail for five decades. She's gutsy and knowledgeable and humorous, most of all she's real—a unique woman of strength and character! Her keen observations and warm sensitivity recreate memorable episodes and touching moments from her years on the trail.

Hardcover, 208 pages, $16.95

Distant Fires
Scott Anderson

A classic canoe-trip story, with a twist of wry. Anderson's journey began on a front porch in Duluth, Minnesota, and ended three months and 1,700 miles later at historic York Factory in Hudson Bay. The reader is treated to a breath of fresh northwoods air with every turn of the page.

Softcover, 176 pages, $12.95

To order, write or call:

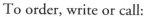

Pfeifer-Hamilton Publishers
210 West Michigan
Duluth MN 55802-1908

Toll Free 800-247-6789
Fax 218-727-0505
Local 218-727-0500